T0338269

Corrosion and Materials in Hydrocarbon Production

Corrosion and Materials in Hydrocarbon Production

A Compendium of Operational and Engineering Aspects

Bijan Kermani

NACE Fellow (F NACE), Fellow of Institute of Materials,
Minerals and Mining (FIMMM), Fellow of Institute of Corrosion (FICorr)
Chartered Engineer (CEng)

Don Harrop

Honorary Fellow of the UK Institute of Corrosion (FICorr)
Honorary Fellow of the European Federation of Corrosion (EFC)

This Work is a co-publication between ASME Press and John Wiley & Sons Ltd.

© 2019 John Wiley & Sons Ltd
This Work is a co-publication between John Wiley & Sons Ltd and ASME Press

The right of Bijan Kermani and Don Harrop to be identified as the authors of this work has been asserted in accordance with law.

Registered Offices
John Wiley & Sons, Inc., 111 River Street, Hoboken, NJ 07030, USA
John Wiley & Sons Ltd, The Atrium, Southern Gate, Chichester, West Sussex, PO19 8SQ, UK

Editorial Office
The Atrium, Southern Gate, Chichester, West Sussex, PO19 8SQ, UK

For details of our global editorial offices, customer services, and more information about Wiley products visit us at www.wiley.com.

Wiley also publishes its books in a variety of electronic formats and by print-on-demand. Some content that appears in standard print versions of this book may not be available in other formats.

Library of Congress Cataloging-in-Publication data applied for

ISBN: 9781119515722

A catalogue record for this book is available from the British Library.

Cover Design: Wiley
Cover Images: © ggw/Shutterstock, © chain45154/Getty Images, © spooh/Getty Images

Set in 10/12pt Warnock by SPi Global, Pondicherry, India

Printed and bound by CPI Group (UK) Ltd, Croydon, CR0 4YY

10 9 8 7 6 5 4 3 2 1

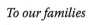

Contents

Preface

Government policy in the pursuit of a carbon-neutral world economy has been a commitment adopted by an increasing number of industrial nations. However, for the foreseeable future, fossil fuels in which hydrocarbons will play a significant role are likely to remain the primary source of energy.

There is a continuing drive to increase the life of existing oil and gas field developments through a number of avenues. These include, for example, tie-backs from nearby smaller reservoirs, otherwise uneconomic to develop alone, and by increasing the recovery rate from existing fields. The average commonly reported recovery rates after primary (natural flow under existing reservoir pressure) and secondary (e.g. water injection, hydraulic fracking) enhancement operations for oil are between 35% and 45%. This could potentially be increased in certain reservoirs by a further 5–15% through tertiary recovery methods (e.g. reducing reservoir oil viscosity by CO_2 flooding, steam, or surfactant injection). Nevertheless, there remains a continuing search for new economic sources of hydrocarbons, taking exploration activities into harsher environments through deep high pressure/high temperature (HPHT) wells, into geographically remote and/or increasing environmentally sensitive areas and deep water. This enterprise has created increased challenges: (i) to the economics of project development and field operations; (ii) on the performance envelope of existing oilfield technology; and (iii) in meeting Health, Safety and Environment (HS&E) commitments which can impact the Licence to Operate (LTO). Ensuring the mechanical integrity of facilities is therefore paramount. The accurate prediction of materials' performance and their optimised selection in tandem with pro-active corrosion mitigation are primary considerations at design and throughout a field's operating life.

Hydrocarbons-producing facilities and infrastructures are potentially subject to both external and internal corrosion threats; in the case of the former from hostile and geographically remote operating environments, and in the latter from the presence of wet produced fluids and acid gases. Both these threat types impact materials selection, engineering design, and through life integrity management (IM).

Corrosion in its various forms remains a major potential threat to successful hydrocarbons production and its optimum control and management are essential to the cost-effective design of facilities and their safe operations. Its impact can be viewed in terms of effect on capital and operational expenditure (CAPEX and OPEX) and HS&E and associated process safety risks. It is, therefore, essential to have a sound corrosion design and management philosophy for production facilities to safely handle and transport wet hydrocarbons enabling integrity assurance and trouble-free operations. Such a philosophy can be used in the technical/commercial assessment of new field development and in prospect evaluation, to prolong the life of ageing assets and, for handling sour fluids by facilities not normally designed for sour service. The book sets out to provide such a philosophy in a pragmatic manner.

The book is intended to be suitable for both practising materials and corrosion engineers working in hydrocarbons production as well as those entering the area who may not be fully familiar with the subject. It is not a textbook; rather it is a practical manual/ready reference source to steer design and operations engineers to currently established best practice drawn on the many years' global experience of the book's authors and contributors. It embodies over 500 years of cumulative field and engineering experience.

The primary focus is on operational and engineering aspects by capturing the current understanding of corrosion processes in upstream operations and providing an overview of the parameters and measures needed for optimum design of facilities. Emphasis is placed on material optimisation which is structured by presenting user-friendly roadmaps. The book is intended to act as an applied tool focusing on engineering features of corrosion and materials.

Chapters on internal corrosion address: the types and morphology of corrosion damage; the principal metallic materials deployed; and mitigating measures to optimise its occurrence. Chapters on external corrosion address corrosion under insulation (CUI), external coating systems and cathodic protection (CP). In addition, a chapter has been assigned to systematically quantifying the level of in-service risk of corrosion, presented in terms of likelihood and consequence, in order to prioritise operational risk. Together with a broader overview of corrosion and integrity management, outlined is a structured and performance-managed approach to the provision of safe and trouble-free operations through an integrated cross-discipline methodology and approach. This is an integral part of meeting compliance with HS&E requirements and legislation and risk management: a primary purpose behind the broader remit of IM.

The book captures and provides solutions via four principal avenues for upstream hydrocarbon operations from reservoir to the refinery and petrochemical plants:

1) Outlining key corrosion threats, both internal and external, and means of inspection, monitoring, control and management.
2) Providing necessary background on types and nature of materials used for the construction of CAPEX-intensive facilities.
3) Underlining current and future challenges that the industry sector is facing with some steer towards respective management and technical solutions.
4) Implementation of effective and progressive materials optimisation, corrosion mitigation methods and corrosion and integrity management strategy.

The final chapter considers the future outlook in energy demand and supply, translating these into technology challenges facing the hydrocarbon production industry sector, which in turn shapes the materials and corrosion technology themes necessary to deliver business success and continuously improve safety, security, and minimise impact on the environment.

It should be noted that there is never a single answer to a potential challenge. The solution may invariably be drawn from a number of options, the convergence of which can lead to an optimum outcome. It is against this background that the book is compiled, allowing flexibility in choice having considered all *credible* corrosion threats and their respective mitigation. The importance of failure analysis in allowing lessons to be learnt is highlighted, together with the importance of in-house, national and international standards in effective implementation of corrosion management and strategy.

Acknowledgement

The book represents much more than merely a compilation of individual experience and thinking on selected topics. The authors would like to express their appreciations to all who have contributed their time and effort to ensuring its successful completion. The authors wish to acknowledge the significant inputs from:

Dr Michel Bonis (France) and particularly in relation to the chapter on Microbial Corrosion.

Professor Bob C. Cochrane (UK) and particularly in relation to the chapter on Carbon and Low Alloy Steels.

Dr Arne Dugstad (Norway) and particularly in relation to the chapter on Corrosion in Dense CO_2 Systems.

Dr David Fairhurst (UK) and particularly in relation to the chapter on Cathodic Protection.

Dr Steve Groves (UK) and particularly in relation to the chapter on Non-metallic Materials and Elastomers.

Dr Paul Hammonds (India) and particularly in relation to the chapters on Corrosion Inhibition and Water Chemistry.

Dr Ian McCracken (UK) and particularly in relation to the chapter on Water Chemistry.

Dr Bill Hedges (UK) and particularly in relation to the chapter on Corrosion Trending.

Mr Roger Howard (UK) and particularly in relation to the chapter on Carbon and Low Alloy Steels

Dr Ali Moosavi (Kuwait) and particularly in relation to the chapter on Coating Systems.

Dr Stefan Winnik (UK) and particularly in relation to the chapter on Corrosion Under Insulation.

Contributions from a number of individuals in relation to their steer in setting the focus and comments on specific chapters or elements are very much appreciated. These include Dr Ed Wade (UK) and Dr Bruce Craig (USA). Notably, Dr Wade made positive and constructive comments and corrective suggestions on many themes considered vital in moving forward. And contributions from Professor Ali Mosleh (USA) and Mr Mehdi Askari (Iran) in reviewing selective chapters and their valuable comments are acknowledged. Provision of a few photographs by Mr Matt Dabiri (US) is highly appreciated.

Particular thanks go to Professor Vic Ashworth (UK) who meticulously reviewed all the chapters and made significant and valuable comments – his efforts and contributions are greatly appreciated. Finally, the wide-ranging inputs from Dr Jean-Louis Crolet (France) throughout the development and completion of the book are acknowledged and appreciated.

Bijan Kermani
Don Harrop

1

Introduction

In the search for new sources of oil and gas, operational activities have moved to harsher environments in deeper high pressure/high temperature wells, remote areas, and deep-water regions. These have created increased challenges for the economy of project development and subsequent operations whereby the integrity of the facilities, optimisation of the materials, and accurate prediction of the materials' performance are becoming paramount. In addition, the economic moves towards multi-phase transportation through sub-sea completions and long infield flowlines have a tendency to increase the risk of corrosion threats, thus placing a heavier duty on integrity management in upstream operations.

Corrosion potentially presents many threats, in many forms, and remains a major operational obstacle to successful hydrocarbon production. These threats have wide-ranging implications for the integrity of many materials used in the upstream petroleum industry, thus affecting capital expenditure (CAPEX) and operating expenditure (OPEX), with consequences on health, safety, and the environment (HS&E). Furthermore, if not effectively identified and managed, in extreme cases, corrosion may have major business implications, such as disruption to production, financial penalties, adverse societal publicity, and even an impact on the Licence to Operate (LTO). However, corrosion mitigation and control by measures through national and international corrosion communities have led to significant improvements in the provision of safety and security and enhancing public welfare.

This chapter sets out to outline three subject areas with a common thread of describing the content of the book and its scope in relation to upstream hydrocarbon production. The subject areas are:

1) the impact of corrosion, highlighting its economic implications;
2) types of corrosion threats in oil and gas production and transportation, the manner in which they manifest, and the means of their control and design;
3) where future priorities need to be set to sustain and develop the continuing fitness-for-purpose of the practice and status of the corrosion and materials discipline.

In addition, brief reference is made to the image of the potential corrosion disciple with a view to outlining future priorities to attract a new generation of high calibre professionals to this field.

Corrosion and Materials in Hydrocarbon Production: A Compendium of Operational and Engineering Aspects, First Edition. Bijan Kermani and Don Harrop.
© 2019 John Wiley & Sons Ltd. This Work is a co-publication between John Wiley & Sons Ltd and ASME Press.

1.1 Scope and Objectives

This book aims to produce a practically driven reference guide to assist corrosion and materials engineers in their quest to select and optimise the most appropriate and economical choice of material and corrosion control strategy for upstream operations.[1] It covers measures and mitigation methods to address corrosion threats in hydrocarbon production systems carrying hydrocarbons, injection water, and/or produced water.

In particular, the book provides an understanding of the primary subject areas that affect the continued and trouble-free operation of hydrocarbon production facilities. It provides a compendium of the principal considerations, current best practice, and key issues associated with each theme without going into absolute detail which will be specific to each individual application.

The focus primarily is on the following topics:

1) Corrosion threats and their respective assessment practices and mitigation methods.
2) Corrosion interrogation methods, including monitoring and inspection data capture and full analysis.
3) Methods by which materials are selected for a particular application.
4) Determining corrosion risk and implications with respect to defining safe operational conditions and the implementation of mitigation methods, measures and practice as an integral part of a fit-for-purpose corrosion and integrity management strategy.
5) Consideration of current and future challenges to those engineers who wish to specialise in materials and corrosion knowledge, and outlining the gaps in best implementation of know-how and knowledge.

While the majority of subject areas relate to addressing internal corrosion, cases of coatings, corrosion under insulation (CUI), cathodic protection (CP), and corrosion trending, combining data generated from corrosion monitoring and inspection, are also included to complement mitigation methods.

The book is intended for use by both competent engineering personnel working in upstream production operations who have knowledge and experience of dealing with corrosion and materials as well as those entering the area who may not be fully familiar with the subject.

1.1.1 Contents of the Book

A summary of the themes and subject areas covered in this book is presented in Figure 1.1.

1.2 The Impact of Corrosion

The impact of corrosion can be viewed in terms of its effect on CAPEX, OPEX, and HS&E. In the past few decades there have been significant studies in various parts of the world on the cost of corrosion and how it affects a country's economy.

According to the current US corrosion study, the direct cost of metallic corrosion is $276 billion on an annual basis. This represents 3.1% of the US gross domestic product (GDP) [1].

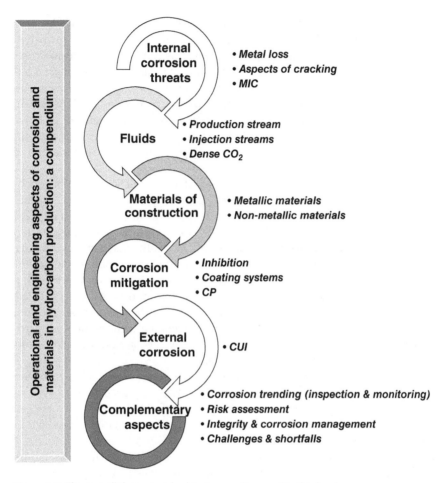

Figure 1.1 The overall themes and subject areas discussed in this book.

The 2016 IMPACT [1] study, released by NACE International, indicates that there are problems with using the existing studies to examine savings over time due to the implementation of corrosion control practices. For instance, in the US, the cost of corrosion was estimated to be equivalent to 2.5% of GDP in 1949 (using the Uhlig method), 4.5% of GDP in 1975 (using the input/output method), and 3.1% of GDP in 1998 (using the Hoar method). The problem is that, in general, these studies use different analyses to estimate the cost of corrosion, so a direct comparison is not possible. Nevertheless, the overall cost of corrosion has been estimated to be between 2% and 5% of GDP, depending on the region of the world, which may be due to differences in methodology. Irrespective of the overall economic impact, the important point is that the cost of corrosion has not actually changed with time, knowledge, or technology since it was first looked at by Uhlig. Therefore, it needs to be reiterated that the use of economic impact alone has limited influence on the interest in raising the importance and funding of corrosion with management.

Domestic oil and gas production is considered a stagnant industry in the US because most of the significant available reserves have been exploited – although the growth in shale gas exploration and production may well change this view. Direct corrosion costs associated with conventional production activity were determined to be about $1.4 billion, with $0.6 billion attributed to surface piping and facility costs, $0.5 billion to downhole tubing, and $0.3 billion to CAPEXs related to corrosion [1]. The NACE 2016 IMPACT [1] study provides valuable tools for companies to implement an effective Corrosion Management System Framework, benchmark their current practices with other organisations worldwide, and learn how to optimise the safety and lifetime of critical assets.

While the overall cost impact of corrosion measured against GDP provides a benchmark for all industrial sectors, a measure focused specifically on hydrocarbon production was needed. To do this, many studies have focused on making the overall cost impact specific and tangible in terms of lifting cost per barrel equivalent. These studies have come up with a consensus view that corrosion failures, the majority of which are related to metal loss CO_2 and H_2S corrosion and cracking threats, account for some 25% of all safety incidents – affecting 2.8% of turnover and 2.2% of tangible assets – resulting in a 8.5% increase on CAPEX, 5% of lost/deferred production, and 11.5% increase to the lifting costs. These are estimated figures and dependent on the operator and region, obtained from a number of publications [2]. They are estimated as the additional corrosion management costs necessary to successfully deploy carbon and low alloy steels (CLASs) as the appropriate construction materials.

The spread of these figures is highly dependent on the manner in which a corrosion control philosophy is planned and implemented, as they vary according to type of operation, location, and operator. The estimated cost of corrosion is put between US$0.3–0.9 (or even higher depending on the conditions and region) for the production of each barrel of oil equivalent (boe) [1–4].

It is outlined by the 2016 IMPACT study that some 10–30% of this cost can be reduced by implementing currently available corrosion control best practices [1]. However, the overall financial impact continues to place heavy penalties despite concerted efforts by the corrosion and materials community. This is primarily due to the increasing move to harsher production conditions and the extended use of CLASs beyond what was previously considered feasible. In addition, while limited, some costly failures have occurred, mainly due to lack of understanding of anticipated exposure conditions or inadequate metallurgical treatment of components, and are discussed in Chapter 19.

1.2.1 The Overall Financial Impact

Several publications are available, describing the financial impact of corrosion in detail. Putting these in context, it can be estimated that with a daily global hydrocarbon production at around 90 million barrels (bbl) equivalent and an average lifting cost of around US$10 US/bbl, the estimated OPEX due to corrosion threats in hydrocarbon production worldwide is around US$103 million dollars per day or over US$38 billion annually [2–4]. The contribution of corrosion threat to downhole operations is estimated at 1.6% of OPEX, making the overall downhole corrosion financial impact some US$5.3 billion annually. These figures, albeit an estimate and dependent on operator/

Table 1.1 Approximate economic impact of corrosion.[a]

OPEX		CAPEX	
11.5% Increase		8.5% Increase	
Activity	Financial impact (¢/bbl)	Activity	Financial impact (¢/bbl)
Downhole	1.6% Overall cost	Corrosion allowance	4
Maintenance	19.6	13%Cr for downhole	4.2
Shutdowns	1.2	Coatings	7.6
Support/inspection	7	Cathodic protection	3.2
Chemicals	5.2		
Major failures	6		
Personnel	0.02		

[a] Figures are based on year 2000 and have taken on board an annual inflation rate of 4–5%. They are approximate in cents/barrels and should be taken as purely indicative as they depend on application, location, operator, and logistics.
For the abbreviations, refer to the Appendix.

region/location, highlight the economic significance of corrosion and the vital role of mitigation, control, and prevention all requiring due and continuous attention. Based on the projection of figures from several publications and taking on board an average annual inflation rate of 4–5% since their publication, a breakdown of this figure as an indication of the economic impact can be summarised in Table 1.1 [2–4]. The increase in OPEX and CAPEX is due to the additional requirement for the implementation of corrosion mitigation measures when using CLAS as the base case. As an example, based on approximate figures in Table 1.1, for producing 90bbl/day, the overall annual cost of chemicals to mitigate corrosion is over US$1.7 billion.

1.3 Principal Types of Corrosion in Hydrocarbon Production

Many reviews and articles have focused on outlining the principal types of corrosion threat in hydrocarbon production systems with a view to channelling attention to key mitigation methods to minimise their occurrence. The reviews have demonstrated the significance of several types of metal-loss corrosion and cracking as the primary types of damage facing the industry [5, 6].

This section summarises the principal types of corrosion threat experienced in hydrocarbon production. This is as a precursor to the subsequent chapters that attempt to deal with individual subject areas in particular detail with respective mitigation methods. By no means does this section aim to describe different modes of corrosion threat as these are classical types and are shown and discussed in many publications.

The risk of internal or external corrosion becomes real once an aqueous phase is present and able to contact the pipe wall, providing a ready electrolyte for corrosion reactions to occur. The inherent corrosivity of this aqueous phase is then heavily dependent on the construction materials, the environmental conditions (temperature, pressure, presence of bacteria, etc.), and levels of dissolved corrosive species (acid gases, oxygen, organic acids, etc.) which may be present.

In hydrocarbon production, corrosion threats are mostly associated with the use of CLAS which continues to be the principal construction material. CLAS has been used extensively mainly due to its excellent properties, versatility, availability, and low cost. However, its inherent corrosion resistance in contact with production and water injection conditions is inadequate if not mitigated and this is the main source of the corrosion threats affecting the design and operation of production and water injection systems.

1.3.1 Corrosion Threats

As referred to earlier, given the conditions associated with oil and gas production and transportation and that of gas and water injection, corrosion must always be seen as a potential risk [7]. Therefore, the need to reliably handle wet hydrocarbons arises from the increasing number of fields where significant levels of CO_2 and H_2S are present under more arduous operating conditions. In addition, the need for increased production which invariably entails water or gas injection to maintain reservoir pressure or enhance sweep/capture of hydrocarbons (cf. recovery efficiency) can introduce O_2 and the potential for microbiological activity introduces further types of corrosion threat.

Extensive industry reviews have shown that corrosion in hydrocarbon production can manifest in several forms, all subject to the prevailing cathodic reaction, i.e. prevailing dissolved cathodic species (e.g. CO_2, H_2S, O_2) driving the overall corrosion reaction [5–8]. While most classical forms of corrosion are encountered in hydrocarbon production, the principal types where the majority of failures occur remain limited. The most prevalent types of damage encountered include metal-loss corrosion and localised corrosion manifested in the presence of CO_2 (sweet corrosion) and H_2S (sour corrosion), dissolved in the produced fluids and by the presence of dissolved oxygen in water-injection systems. These three types of corrosion threat are each addressed specifically in the present book due to their importance in terms of frequency of occurrence and the respective cost impact they impose on both CAPEX and OPEX [1–4].

This book deals primarily with aspects of internal corrosion, including microbiologically induced corrosion (MIC). However, as CUI continues to pose operational challenges in its detection and mitigation, Chapter 13 is allocated to addressing this type of external corrosion threat. Also Chapter 9 covers the challenges addressed by the coatings industry in controlling primarily external corrosion alone and in the presence of cathodic protection, where applicable.

Forms of corrosion of CLASs exposed to hydrocarbon production conditions are summarised schematically in Figure 1.2. Some of these are specific to an alloy/environment system and may not necessarily affect corrosion-resistant alloys (CRAs). The majority of these threat types are dealt with in separate chapters. It was not considered necessary to describe in classical detail each of the generic types of corrosion shown in Figure 1.2, as this is covered in most standard corrosion textbooks.

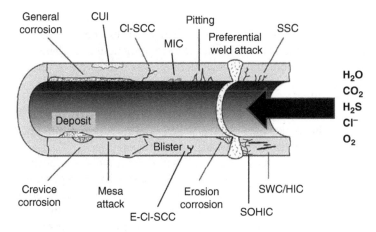

Figure 1.2 Principal types of corrosion threats on CLAS in hydrocarbon production.

1.4 The Way Ahead: Positive Corrosion

While the focus of this text is to describe practical measures to deal with corrosion in upstream operations, it is necessary to mention a few notes on the future role of the corrosion and materials discipline so that it can continue to successfully address corrosion threats.

It should be emphasised that the corrosion community has contributed enormously to the advances made in corrosion science, technology, and engineering which have subsequently led to step changes in material degradation mitigation practices [8]. The global awareness of the topic, including succinct communication, education, and public awareness, has come a long way over the past few decades. The impact of such measures has been felt across nations, governments, communities, and cultures. The corrosion profession across academe and industry, underpinned by professional bodies such as NACE International and the European Federation of Corrosion (EFC) and smaller national societies, continues to have substantial impact, with encouraging progress.

Corrosion remains a very interesting, challenging, exciting, rewarding, and relevant subject incorporating a diverse set of disciplines: Physics, Metallurgy, Chemistry, Engineering, and Art. Nevertheless the corrosion discipline is often in itself insufficiently always ready to intake and retain high calibre personnel. This is a critical loss to the industry and it is imperative that the anticipated shortfall is addressed effectively and sustainably. Furthermore, the discipline average age is increasing with the decreasing number of young engineers and scientists entering the community. According to NACE International, more than 60% of the members are aged above 40 and more than 40% aged above 50 years old. This deficit is believed to be correlated to the current niche image of corrosion within the professional society [8].

Reducing the number of dangerous events, injuries, and undesirable releases remains a top priority and the key focus of the profession's commitment to continually improving industrial and social safety standards. Corrosion societies across the globe have maintained a relentless effort to ensure that safety performances is improved through facilitating leadership, communication, education, and technology transfer. Looking to the future, the discipline will continue to be a key player in enabling the more efficient use of resources, thus reducing the impact on climate change. For example, the corrosion community has an instrumental role to play in the development and implementation of carbon capture, transportation, and storage technologies that can help reverse the effect of CO_2 emissions.

While the economic and HS&E impacts of corrosion and proactive prevention of corrosion-related failures are significant and major drivers, it still appears predominantly to be only the failures that make the headlines beyond the corrosion community, thus further underpinning a somewhat negative image of corrosion.

It is crucial to advance a different image of the corrosion community – a 'positive image' [8]. To underline what the corrosion community is capable of doing and has done in facilitating environmental benefits, help secure and facilitate wealth creation and the provision of societal well-being, safety and security. In sustaining the future of the community, it must attract a new generation of young enthusiastic engineers to the discipline, and it will be much more successful by presenting a positive vision rather than a 'life insurance' perspective.

1.5 Summary

- Corrosion failures account for 25% of all safety incidents, 2.8% turnover, 2.2% tangible asset, 8.5% increase on CAPEX, 5% of lost/deferred production, and 11.5% increase to the lifting costs, or more quantitatively between US$0.3–0.9/boe. OPEX uplift is governed by the operating regime and location, the maturity of the asset, and other parameters.
- Principal forms of corrosion threat in hydrocarbon production were outlined to include those in relation to the presence of H_2S and CO_2 in produced fluid and O_2 in injected fluids.
- The majority of corrosion threats in upstream oil and gas production are associated with metal loss CO_2 corrosion of CLASs (with implications for CAPEX and OPEX and HS&E) and are operator-dependent.
- The industry continues to lean heavily on the extended use of CLASs which are readily available and able to meet many of the mechanical, structural, fabrication, and cost requirements. Their technology is well developed and they represent for many applications an economic materials choice. However, a key obstacle – their Achilles' heel – to their effective use is their limited corrosion performance.
- This book provides an overview of the principal considerations and key issues associated with hydrocarbon production and transportation – a compendium of current best practices and state-of-the-art knowledge presented by expert and highly experienced field practitioners – to identify credible corrosion threats and their safe and cost-effective mitigation and proactive management. The content is meant to provide flexibility to be used in conjunction with competent technical judgements. Nevertheless, it remains the responsibility of the end user to judge its specific relevance and suitability for a particular application or context.
- Bearing in mind the influential role that the corrosion and materials discipline has played in providing improved safety and security, the community needs to be portrayed and championed in a positive way to attract a younger generation of high calibre individuals to further assist in the task of dealing with corrosion problems.

Note

1 Upstream production operation is defined as the facilities and equipment involved in the extraction, handling/processing and transportation of hydrocarbon fluids and gases from reservoirs, via well or production sites to central process sites and then finally to refinery or petrochemical plants. It excludes facilities in relation to refineries or petrochemical plants.

References

1 NACE, International measures of prevention application, and economics of corrosion technologies study (IMPACT), NACE International, Report No. OAPUS310GKOCH (PP110272)-1, 2016.
2 Kermani, M.B. and Harrop, D. (1996). The impact of corrosion on the oil and gas industry. *SPE Production & Facilities* 11 (3): 186–190.
3 McIntyre, P. (2002). Cost of corrosion and the role of corrosion standards. *Corrosion Management* 46: 19–21.
4 M Bonis and P Thiam, Measurement of corrosion costs within Elf's exploration and production. Paper presented at Eurocorr, 1999.
5 Craig, B. (2004). *Oilfield Metallurgy and Corrosion*, 3e. Washington, DC: NACE International.
6 Jones, L.W. (1988). *Corrosion and Water Technology for Petroleum Producers*. Tulsa, OK: OGCI Publications.
7 Kermani, M.B. and Morshed, A. (2003). CO_2 corrosion in oil and gas production: a compendium. *Corrosion* 59: 659–683.
8 Kermani, M.B. (2014). Positive corrosion. *Materials Performance* 53: 14–20.

Bibliography

Atkinson, V.D. (1995). *Corrosion and Its Control*, 2e. Houston, TX: NACE International.
Bradley, R, Oil and gas isn't just one of the richest industries, it's also one of the safest. Forbes, March 25, 2013.
Energy Institute (2010). *Technical Guidance on Hazard Analysis for Onshore Carbon Capture Installation and Onshore Pipelines: A Guidance Document*, 1e. London: Energy Institute.
ExxonMobil Corporation, The outlook for energy: a view to 2040. 2017. www.corporate.exxonmobil.com/.../energy-outlook/a-view-to-2040 (accessed 23 September 2018).
Fontana, M.G. (1986). *Corrosion Engineering*, 3e. New York: McGraw-Hill.
International Association of Oil and Gas Producers, Environmental performance indicators, 2011 data. OGP Report No. 2011e, October 2012.
International Association of Oil and Gas Producers, Risk assessment data directory: riser and pipeline release frequencies. Report No. 434-4, March 2010.
International Tankers Owners Pollution Federation (2016). *Oil Tanker Spill Statistics*. London: ITOPF www.itopf.org/fileadmin/data/Documents/Company_Lit/Oil_Spill_Stats_2016.pdf.

Kermani, MB, Materials optimisation for oil and gas sour production. NACE Annual Corrosion Conference, Paper No. 00156, 2000.

Kermani, M.B. (ed.) (2013). *Carbon Capture, Transportation and Storage (CCTS): Aspects of Corrosion and Materials*. Houston, TX: NACE International.

Koch, G, Varney, J, Thompson, N, et al., International measures of prevention, application, and economics of corrosion technologies study (IMPACT), 2016. NACE Report. www.impact.nace.org/documents/Nace-International-Report.pdf (accessed 3 October 2018).

Manhattan Institute for Policy Research, Issue brief. No. 23, June 2013.

National Materials Advisory Board (2009). *Assessment of Corrosion Education*. Washington, DC: National Academy of Sciences.

National Materials Advisory Board (2011). *Research Opportunities in Corrosion Science and Engineering*. Washington, DC: National Academy of Sciences,.

National Transportation Safety Board, NTSB accidents and accident rates by NTSB Classification, 1998–2007, 2008. https://catalog.data.gov/dataset?publisher=National Transportation Safety Board (accessed 3 October 2018).

Nuclear Energy Institute, US nuclear industrial safety accident rate, 2012, www.nei.org/resources/statistics (accessed 3 October 2018).

Shell E & P Ireland Ltd, Corrib onshore pipeline QRA, DEN, Report No./DNV Reg. NO.: 01/12LKQW5–2, Rev. 01. 18 May 2010.

Shreir, L.L., KoJarman, R.A., and Burstein, G.T. (eds.) (1994). *Corrosion*, 3e. Oxford: Butterworth Heinemann.

U.S. Department of Transportation, Pipeline and Hazardous Materials Safety Administration Office of Pipeline Safety (PHMSA), Integrity management program. www.transportation.gov/.../integrity-management-program (accessed 3 October 2018).

2

Carbon and Low Alloy Steels (CLASs)

Irrespective of their intended use, steel products are an essential enabling commodity for the oil and gas industry, forming the backbone of construction. Among commonly used steel products, carbon and low alloy steels (CLASs) constitute the primary type of steel used in the hydrocarbons production industry sector.

CLASs are a category of ferrous materials that offer superior mechanical properties to plain carbon steels as the result of the addition of alloying elements, such as nickel, chromium, and molybdenum. Total alloy content can range from around 2% up to levels just below that of stainless steels, which typically contain a minimum of 12%Cr. The primary function of the alloying elements is to increase hardenability[1] in order to optimise the mechanical properties and toughness[2] after heat treatment.

The oil and gas industry sector continues to lean heavily on the use of these steels which are readily available in the volumes required and able to meet many of the mechanical, structural, fabrication, and cost requirements. Their technology is sufficiently developed and for many applications these materials represent an economical choice. However, the inherent corrosion resistance of CLASs is relatively low. Consequently, their successful application invariably requires one or more whole-life forms of corrosion mitigation, against both internal and external exposure conditions. Corrosion mitigation options are addressed in other chapters and therefore not considered here with the exception of the use of internal corrosion resistant alloy (CRA) cladding.

This chapter aims to provide an overview of the CLASs used in hydrocarbons production. However, it is not intended to provide detailed metallurgical and production parameters. Rather, it is intended to serve as a reference to introduce the subject for those not familiar with metallurgy and steel production. Detailed information on metallurgical engineering can be found in classic literature, including those identified in the Bibliography.

2.1 Steel Products

A wide variety of steel grades are used across the industry with the most notable examples briefly described in this section.

Corrosion and Materials in Hydrocarbon Production: A Compendium of Operational and Engineering Aspects, First Edition. Bijan Kermani and Don Harrop.

2.1.1 Structural Services

Support structures commonly use 'structural' steel products in the form of rolled plate, various sectional shapes, and tubulars, although cast or forged products may also be used. Typical applications include offshore structures, sub-sea module support frames, pipe racks, process equipment saddles, and, in some instances, low criticality storage tanks.

For structural purposes, carbon steels are frequently used. Carbon steels invariably contain Mn (typically 0.5–1.5%) as a strengthening element and limited quantities of other alloying elements such as Nb, Ti and V are often employed, producing so-called 'lean alloy or micro-alloyed steels' in order to achieve the desired range of mechanical properties.

Structural steels generally require adequate properties over a range of temperatures lying within the limits of approximately –50 °C to +50 °C, depending on the local environmental conditions, e.g. in Arctic, temperate, or tropical conditions.

2.1.2 Pressure Containment

C/Mn or lean alloy steels may be used for process plant, vessels, pipework, pipe fittings, and valve bodies requiring pressure containment. However, steels with an increased alloy content of Cr, Ni, Mo, the so-called low alloy steels, are employed to produce mechanical properties suitable for a temperature range lying within the limits of approximately –80 °C (characteristic of Joule Thompson cooling) to approaching 600 °C (characteristic of a number of refining processes). Typical applications requiring pressure containment include drill pipes, casing and tubing, line pipes, process pipework, pressure vessels, and heat exchangers.

Gas liquefaction and storage applications require steels with good low temperature properties. Down to approximately –50 °C, low temperature C/Mn steels or 2.25% Ni steels may be used. However, at lower temperatures, aluminium alloys are often used in heat exchangers while 3.5–9% Ni steels or austenitic stainless steels are used for storage at progressively lower temperature down to –196 °C.

Steels with substantially higher contents of Cr, Ni, Mo, i.e. stainless steels and other CRAs, are discussed in Chapter 3.

2.2 Development of Mechanical Properties

The mechanical properties of C/Mn, micro-alloyed, and low alloy steels depend on the chemical analysis and the associated microstructure, and are also significantly influenced by heat treatment and mechanical working.

The ability of steel to develop a wide range of properties largely depends on the allotropic transformation of iron. At temperatures above 937 °C, iron has a face-centred cubic crystal structure, known as austenite, while below this temperature, the structure transforms to a body-centred cubic crystal structure and is known as ferrite.

The presence of carbon in a steel leads to the formation of iron carbide (Fe_3C), the proportion of carbide increasing as the carbon content increases. Many of the steels employed for structures and plants in the hydrocarbons production industry contain less than 0.25% C; although, on occasions, low alloy steels with up to 0.40% C are employed. However, for downhole well completion applications, steels having carbon contents in the range 0.20–0.55% are generally employed.

2.2.1 Heat Treatment

Heat treatment assists in the development of optimal microstructures and mechanical properties. On extremely slow cooling from an elevated 'austenitising' temperature (say, 950 °C–1000 °C), for example, in a closed heat treatment furnace with the heating turned off, carbon steel would develop relatively soft microstructures of ferrite and carbide. Such slow cooling allows a microstructural transformation by diffusion to occur to the maximum extent and the so-called 'equilibrium microstructure' is formed. However, industrial processes seldom, if ever, allow for such slow cooling to take place. This section outlines some commonly used heat treatment terminologies.

2.2.1.1 Normalising

In simplistic terms, a still air cool, known as normalising, is usually the slowest cool encountered in practice and while a ferrite/carbide microstructure may still develop, progressively faster cooling rates will not allow the full equilibrium transformation from austenite to ferrite and the formation of Fe_3C to occur.

2.2.1.2 Quenched and Tempering

Rapid cooling, such as a water quench from an austenitising temperature, can suppress the austenite to ferrite transformation since there is no opportunity for diffusion to occur. Rather, a shear transformation of the austenitic crystal structure takes place and a relatively hard brittle acicular microstructure, known as martensite, is formed without the presence of Fe_3C.

The subsequent introduction of a heat treatment at around 600 °C, known as tempering, will allow this martensitic microstructure to break down and carbides to precipitate. A long-term heat treatment at this temperature would again lead to the development of the equilibrium microstructure.

2.2.1.3 Mechanical Working

The influence of mechanical working on the development of mechanical properties is also important. Industrial bulk casting of steel does not generally develop optimum microstructures and mechanical properties, and is invariably followed by some form of mechanical working. For example, the grain size in as-cast steel may be relatively large and this is not generally conducive to optimal properties. Mechanical working by forging or rolling at elevated temperature (approx. 1000–1200 °C) will break up this coarse grain structure and result in a finer grain structure by a process of re-crystallisation[3] resulting in improved mechanical properties, such as yield strength,[4] ductility,[5] and toughness. In practice, mechanical working within specific temperature ranges and the control of the associated cooling rates can lead to a range of microstructures, varying from ferrite and carbide to martensite and also intermediate microstructures containing ferrite, carbide, and other acicular structures known as 'bainitic'.

2.2.2 Industrial Processes

A number of industrial mechanical working/heat treatment process are employed in-order to produce modern steels. Probably the simplest process is mechanical working followed by normalising. Mechanical working followed by quenching and tempering (Q&T) may also be used while other more specialised, and often proprietary, mechanical working/heat treatment/cooling regimes known as thermo-mechanically controlled processes (TMCP) are frequently employed to produce micro-alloyed steels.

Depending on the chemical analysis, normalised steels will typically generate yield strengths up to 415 MPa (grade X60). Similarly, quenching and tempering and TMCP processes may potentially result in yield strengths of the order 690 MPa (grade X100).

2.3 Strengthening Mechanisms

Section 2.2.2 provides an overview of the processes employed to achieve a range of mechanical properties in C/Mn, micro-alloy and low alloy steels, where several strengthening mechanisms are involved. These mechanisms are common to all metallic materials and alloys and involve interaction with dislocations, i.e. crystallographic structural imperfections. Dislocations are formed during solidification and also by the application of strain. Hindrance to dislocation movement results in an increase in yield strength and an associated decrease in ductility.

The strengthening mechanisms include:

1) Solid solution strengthening (substitutional and interstitial).
2) Grain refinement.
3) Mechanical work.
4) Dispersion strengthening.

Typical strengthening processes for different steel products and the range of yield strengths are summarised in Table 2.1.

2.3.1 Solid Solution Strengthening

Substitutional solid solution strengthening occurs when alloy elements substitute for the primary metal atoms in the crystallographic structure. Substitutional atoms result in lattice distortion, thus inhibiting dislocation movement. In the case of steels, Mn, Si, Cr, Ni and Mo additions lead to substitutional solid solution strengthening, in addition to their influence on hardenability.

Interstitial solid solution strengthening occurs when relatively small atoms are located in the crystallographic interstices. In steels, C and also N (generally considered an impurity) lead to interstitial solid solution strengthening. These interstitial atoms readily diffuse through the crystallographic structure at relatively low temperatures and predominantly are found at the dislocations, effectively locking their movement. The phenomenon of strain ageing at temperatures in the range 20–150 °C occurs due to interstitial diffusion.

Table 2.1 Typical strengthening process for metallic materials used in hydrocarbon production.

Steel type	Product	Application	Strengthening methods	Range of nominal yield strength
Carbon and low alloy steels	Flat products	Pipelines, vessels and large diameter piping	Alloying elements Heat treatment Thermo-mechanical Control processing	<100ksi (690 MPa)
	Seamless tubular products	Downhole	Alloying elements Cold working and heat treatment	<150ksi (1030 MPa)
Corrosion-resistant alloys	Seamless tubular products	Downhole	Alloying elements and cold drawing	<140ksi (965 MPa)
	Components	Downhole and others	Alloying elements and mixture hardening	~150ksi (1035 MPa) (in some instances can achieve 200ksi, 1380 MPa)

2.3.2 Grain Refinement

Grain boundaries effectively prevent the movement of dislocations. Under the influence of an applied stress, dislocations pile up at the grain boundaries until a critical stress is reached and a dislocation movement is triggered in the adjacent grain. Thus, the finer the grains, the greater the number of grain boundaries and the greater the strength.

2.3.3 Mechanical Working

Mechanical working may be performed at high or low temperatures. High temperature working, relying on recrystallisation, and controlled processing/phase transformation is outlined above.

Cold working at relatively low temperatures generates high levels of dislocations, resulting in significant hardening and increases in yield and tensile properties.[6]

2.3.4 Dispersion Strengthening

Dispersion strengthening relies on the presence of a second phase, possibly more, within the microstructure. The precipitation of a fine second phase within the microstructure will inhibit the movement of the dislocations and thus result in a strengthening effect. In C/Mn steels, carbides (Fe_3C) generally provide dispersion hardening. In lean alloy steels, the addition of Nb, Ti and V also result in the precipitation of carbides. In low alloy steels, Mo_2C often provides dispersion hardening.

2.4 Hardenability

Increasing the carbon level and, to a lesser extent, the alloy content of a C/Mn steel increases the hardenability of the steel. Hardenability is a measure of the ability of the steel to harden, i.e. form martensitic microstructures, through its entire thickness. Thus, a relatively low hardenability steel, would, for example, form a through thickness martensitic microstructure on water quenching at a maximum thickness of, say, 25 mm. However, a steel exhibiting high hardenability would, for example, form, through thickness, martensitic microstructures on normalising at, say, a thickness of 100 mm. In each case, as noted above, a tempering heat treatment would be necessary to develop optimum mechanical properties.

2.5 Weldability

Weldability is a measure of the ease of welding. In order to have good weldability, a steel would develop optimum mechanical properties in the heat-affected zone (HAZ), i.e. the region immediately adjacent to the weld, in the as-welded condition. The HAZ will be exposed to high temperatures, up to the melting point of the steel, on welding, and then will be subject to rapid cooling. Thus, a relatively low hardenability is required to ensure maximum weldability. However, higher hardenability steels may still be welded by applying a pre-heat to the joint prior to welding and maintaining the pre-heat level until weld completion in order to slow the subsequent cooling rate. In the extreme, a post-weld tempering heat treatment may be applied to develop the optimal HAZ mechanical properties.

The weldability of C/Mn steels may be assessed in terms of carbon equivalent (CE). CE is a means of relating the alloy content of the steel in terms of the carbon content and is based on Eq. (2.1):

$$CE = C + Mn/6 + (Cr + Mo + V)/5 + (Cu + Ni)/15 \tag{2.1}$$

This equation is known as the International Institute of Welding Carbon Equivalent, i.e. CE_{IIW}, and is used for steels containing 0.05–0.25% C, 1.7% Mn max, Cr 0.9% max, Cu 1.0% max, Ni 2.5% max, Mo 0.75% max, V 0.02% max, having a CE_{IIW} value in the range 0.3–0.7.

For modern lean alloy steels having less than a 0.12% C content, an alternative equation is often employed (Eq. 2.2):

$$CE_{Pcm} = C + Si/30 + Mn/20 + Cu/20 + Ni/60 + Cr/20 + Mo/15 + V/10 + 5B \tag{2.2}$$

Each of the above equations demonstrates the overriding influence of carbon on weldability and thus, hardenability.

While low CE_{IIW} steels may be welded without pre-heat or the application of post-weld heat treatment (PWHT), C/Mn steels are potentially susceptible to hydrogen cracking[7] when arc welded and pre-heat may be necessary to mitigate this phenomenon. Standards, such as BS EN 1011-2/AWS D1.1, provide a methodology for the calculation of pre-heat levels for the avoidance of hydrogen cracking related to CE_{IIW} and similar formulae for various welding conditions and joint details. For ease of welding, standards for C/Mn steels and structural steels generally specify a maximum CE_{IIW} of 0.42/0.43 or a CE_{Pcm} of 0.21/0.22.

Low alloy steels invariably require the application of pre-heat and PWHT in order to avoid the possibility of hydrogen cracking due to welding and mitigate the effects of their increased hardenability. Recommended pre- and post-weld heat treatment temperatures are given in most welding standards.

2.6 Line Pipe Steels

While weldability is an important issue for the installation of pipelines, the compositional limits on C/Mn and micro-alloy steels for line pipe depend, to an extent, on the manufacturing process. For example, seamless pipe, having a relatively high thickness-to-diameter ratio, is manufactured from a solid billet by a forging/piercing process, e.g. mandrel rolling, a plug mill or pilger process, which will induce a degree of mechanical work. This working, combined with a thermal treatment, such as normalising or quenching and tempering, will largely determine the mechanical properties and thus higher carbon levels are generally specified than for seam-welded line pipe.

Seam-welded line pipe (with a relatively small thickness-to-diameter ratio) is manufactured from strip or plate (flat product). Manufacturing processes include pressing, e.g. the UOE process[8] or roll forming, followed by seam welding. The strip or plate will have undergone a high degree of mechanical work and, in many instances, specific TMCP processes, allowing the use of lower carbon micro-alloy compositions to achieve the required mechanical properties.

Thus, in principle, lower carbon equivalent values are available for the majority of grades of welded pipe. Nevertheless, in the case of both seamless and welded pipes, maximum CE_{IIW}/CE_{Pcm} values of 0.42–0.43/0.21–0.22 can readily be met by manufacturers for all commonly used grades; however, many company specifications apply further restrictions to ensure maximum weldability.

2.7 Well Completion Downhole Tubulars

Well completions rely on a number of components, including casing and tubing, which are typically seamless products. Downhole completions rely on threaded connections for installation and hence field weldability is not an issue.

As noted above, seamless tubular manufacturing processes typically do not allow the use of thermo-mechanical processing that is frequently employed for the strip or plate used for welded pipe. Therefore, the strength of seamless tubulars is normally obtained by alloying the contents in combination with a suitable heat treatment, such as quenching and tempering.

As weldability is not normally an issue for casing and tubing, relatively high carbon contents, up to 0.55%, depending on the strength grade, manufacturing and processing route, are invariably employed. Additional alloying elements, such as Cr, Ni and Mo, are also employed in some tubing grades to assist in the development of optimum properties.

Another seamless component used in exploration and production is drill pipe. This is a heavy, seamless, high strength, low alloy steel tubular, again coupled with threaded connections.

2.8 Internally Clad Materials

While C/Mn, micro-alloy and low alloy steels are essential materials for the hydrocarbons production industry, the major limitation to their application is their relatively poor corrosion resistance to many of the associated fluids. By contrast, high alloy CRA materials can invariably be used in the range of aggressive environments that are encountered, but the strength level and the whole life cost of the necessary alloying elements are often factors that potentially can restrict their application. However, by applying a high alloy CRA material as a thin (3–5 mm) internal cladding layer to a C/Mn, micro-alloy or low alloy steel base layer, the strength and cost issues can, in many instances, be effectively addressed.

Plate materials can be clad through explosive bonding or roll bonding techniques. The latter process is more common and a wide range of ferritic and austenitic stainless steels, nickel base alloys, and titanium can be applied on top. The clad plates may then be used to manufacture internally clad process vessels or internally clad, longitudinally welded process vessel or line pipe. Both the explosive bonding and roll bonding techniques produce a clad layer that is metallurgically bonded to the steel backing material.

Alternatively, the clad layer may be generated by the use of overlay welding. A number of welding techniques are available for the deposition of CRA cladding, both for large surface areas, such as the inside of process vessels, and smaller surfaces or restricted access areas, such as the internal surfaces of castings and forgings, vessel nozzles, or seamless line pipe. Clearly, the weld deposition of a clad layer will result in the fusion of the CRA with the steel backing, thus forming a metallurgical bond.

A further technique that may be employed for the production of CRA-lined pipe materials involves the use of either mechanical or hydraulic expansion. A CRA liner pipe inserted into a steel carry pipe is internally expanded, creating an extremely tight fit, a mechanical bond, between the two materials. While there is no metallurgical bond between the materials, for many pipeline applications, this may not be a limiting factor.

Typical examples of commonly used CRAs for flowlines and pipelines, together with their typical threats, are given in Table 2.2, offering general guidance, with more details on properties covered in Chapter 3. However, for each specific application, a materials performance evaluation should be conducted to ensure suitability for the intended service.

2.9 Summary

C/Mn, lean alloy, and low alloy steels are the principal materials of construction used in the hydrocarbons production industry. These steels represent an economical choice and are available with a wide range of mechanical properties to meet the requirements of structural applications, downhole completions, low temperature and high temperature process equipment, and pipelines.

Table 2.2 Commonly used CRAs in flowlines and pipelines.[a]

Generic type	Examples	Typical composition (wt%)				Typical practical limitations on use	Typical usage
		Cr	Ni	Mo	Others		
Super martensitic stainless steel	Super 13%Cr (weldable)	11.5–13.5	4.5–6.5	2–3		• Some susceptibility to chloride sulphide stress cracking (SCC) at elevated temperatures • Pitting and crevice corrosion in chloride-containing waters with residual oxygen • SSC under some conditions • Hydrogen cracking under some conditions (watch out for CP).	Used as solid CRA only. Often pipe in pipe
Austenitic stainless steel	316L	18	12	0.5		Chloride stress corrosion cracking (SCC) threat in presence of oxygen and/or H_2S at elevated temperatures Pitting corrosion at lower temperatures	Used as liner, cladding, overlay or solid
Duplex stainless steel	22%Cr Duplex	22	5.5	3	N	SCC risk at higher temperatures. SSC under some conditions Hydrogen cracking under some conditions. Pitting and crevice corrosion in chloride-containing waters with oxygen.	Used as liner, or solid
	25%Cr Superduplex	25	7	3.5	N (Cu, W)	SCC risk at higher temperatures SSC under some conditions Hydrogen cracking under some conditions. Pitting and crevice corrosion in chloride-containing waters with oxygen at higher temperatures.	Used as liner, or solid
Nickel base alloy	Alloy 625	22	60	9	Fe, Nb	Crevice corrosion under some conditions.	Used as liner, cladding or overlay

[a] Reproduced with permission from EFC, [1].

Material processing by heat treatment and mechanical working was briefly reviewed, together with fundamental strengthening mechanisms and the importance of weldability has been highlighted.

Notes

1 Hardenability is the ability of a material to become hardened to a specified depth below its surface, and specifically a measure of the depth to which a material will harden on quenching: the maximum hardness is mainly a function of the carbon content. Hardness is normally defined in engineering as the resistance of a material to mechanical indentation, and is the general indication of strength of a material as well as resistance to wear and scratching.
2 Toughness is the ability of a material to withstand a suddenly applied load and thus absorb a certain amount of energy without failure; it depends upon both the strength and ductility of the material.
3 Re-crystallisation: relatively coarse austenite grains re-crystallise to a finer grain structure, ultimately leading to relatively fine microstructural products on subsequent processing.
4 Yield strength is the stress at which a material begins to deform plastically.
5 Ductility is the extent to which a material can sustain plastic deformation before rupture, the ability to undergo considerable permanent strain or deformation before breaking.
6 Tensile strength is the maximum stress that a material can withstand before failing, often referred to as ultimate tensile strength (UTS).
7 Hydrogen cracking may occur as a result of the welding process through a combination of hydrogen generated by the welding process, a hard brittle structure susceptible to cracking and tensile stresses acting on the welded joint. The principal source of hydrogen is moisture contained in the flux used in submerged arc welding or other significant sources of hydrogen, e.g. from the material, where processing or service history has left the steel with a significant level of hydrogen or moisture from the atmosphere. Hydrogen may also be derived from the surface of the material or the consumable. This hydrogen, when trapped in the matrix, can lead to cracking and occur sometimes after the welding operation is completed.
8 UOE is a process in which flat plate is pressed into a U shape, further pressed into a nominal O shape, seam welded, and then internally expanded to form a length of line pipe.

Reference

1 Kermani, B. and Chevrot, T. (eds.) (2012). *Recommended Practice for Corrosion Management of Pipelines in Oil and Gas Production and Transportation*. The Institute of Materials, European Federation of Corrosion.

Bibliography

API/ISO Specifications

API, Specification for line pipe. API Specification 5L (ISO 3183), 2012.
API, Specification for CRA clad or for lined steel pipe. API Specification 5LD, 2015.
API, Specification for CRA line pipe. API 5LC, 2015.

API, Specification for steel plates for offshore structures, produced by thermo-mechanical control processing (TMCP). API Specification 2W, 2006.

API, Specification for carbon manganese steel plate for offshore structures, API Specification 2H, 2016.

API, Petroleum and natural gas industries – steel pipes for use as casing or tubing for wells. API Specification 5CT (ISO 11960), 2014.

AWS, Structural welding code – steel. AWS D1.1, 2015.

British Standards, Welding: Recommendations for welding of metallic materials. BS EN 1011, Parts 1 and 2, 2009.

British Standards, Specification for unfired fusion welded pressure vessels. BS PD 5500, 2018.

British Standards, Weldable structural steels for fixed offshore structures – technical delivery conditions. EN 10225 2009, 2009.

British Standards Centre, Specification for unfired fusion welded pressure vessels. PD 5500, 2016.

EEMUA, Line pipe specification – clauses in addition to API 5 L / ISO 3183. EEMUA Publication 233, 2016.

ASME Standard

ASME, Welding, brazing, and fusing procedures, welders, braziers and welding brazing and fusing operators, 2017ASME B31.3, Process Piping. ASME BPVC IX, 2017

Further Reading

Ashby, M.F. (2005). *Materials Science in Mechanical Design, 3*. Oxford: Elsevier.

Ashby, M.F. (2017). *Materials Selection in Mechanical Design*, vol. 5. Oxford: Elsevier.

Bhadeshia, H.K.D.H. and Honeycombe, R.W.K. (2017). *Steel Microstructure and Properties, 4*. Oxford: Elsevier.

Brooks, C.R. (1999). *Principles of the Heat Treatment of Plain Carbon and Low Alloy Steels*, vol. 2. ASM International.

Llewellyn, D.T. and Hudd, R.C. (2017). *Steels*, 3e. Oxford: Elsevier.

Reardon, A.C. (2011). *Metallurgy for the Non-Metallurgist*. ASM International.

Verhoeven, J.D. (2007). *Steel Metallurgy for the Non-Metallurgist*. ASM International.

3

Corrosion-Resistant Alloys (CRAs)

Potential materials options and corrosion mitigation methods for the construction of hydrocarbon production facilities are limited and subject to a number of considerations. While carbon and low alloy steels (CLASs) alone or in combination with corrosion prevention systems may be suitable for some applications, another category of alloys, containing varying principal alloying elements including Fe, Cr, Ni and Mo, has emerged as an alternative choice. Generally, these fall within the category of corrosion-resistant alloys (CRAs). They may be economic alternatives to CLASs when metal-loss corrosion makes the latter unreliable for the required service life.

Even though CRAs are more costly to procure in terms of CAPEX, they may offer more favourable whole life cost as they lower the risk in terms of corrosion threats. In addition, some applications, particularly in corrosive high-pressure/high-temperature (HPHT) wells often demand the application of CRAs as the only means to control corrosion and allow production from deep hot reservoirs.

Since the early applications of CRAs for petroleum production in wells, the use of these alloys has also been extended to facilities equipment, flowlines, and pipelines. In these applications CRAs are often applied as liners and cladding within a CLAS carcass for economic reasons. In addition, solid stainless steel pipelines and, in a few cases, solid nickel alloy pipelines have also been used for flowlines and pipelines applications.

This chapter introduces the categories of CRAs commonly used in the industry, the important metallurgical factors characteristic of the general groups of CRAs and the typical limits to application of these alloys. In addition, a check list aiding selection through qualification and quality control is provided. A brief overview of notable points to consider when considering well completion is also included.

3.1 Background

Depending on the grade of CRA, their metal loss corrosion is generally insignificant due to the addition of different elements, such as Cr, Ni, and Mo, as a result of the formation of a persistent surface passive layer. Limits of the application of CRAs normally rest on the possibility of corrosion threat by environmental cracking (EC) or localised corrosion [1–3]. CRA resistance to EC and to some extent localised corrosion is only satisfactory if it meets the requirements of ISO15156/NACE MR0175 [1]. Nevertheless, it is

Corrosion and Materials in Hydrocarbon Production: A Compendium of Operational and Engineering Aspects, First Edition. Bijan Kermani and Don Harrop.
© 2019 John Wiley & Sons Ltd. This Work is a co-publication between John Wiley & Sons Ltd and ASME Press.

important to note that certain CRAs can be susceptible to hydrogen-assisted cracking (HAC) as a result of welds exposed to cathodic protection (CP). Likewise, CRA subsea bolting has been found to be very susceptible to HAC when exposed to CP.

A steel is referred to as 'high alloy' if it contains at least 5% alloying elements. Properties which characterise these steels are:

- corrosion resistance;
- high temperature resistance;
- heat and scale resistance;
- low-temperature impact resistance.

This chapter only covers the addition of elements in relation to improving corrosion resistance and the so-called development of CRAs. In relation to resistance to erosion which may be directly related to hardness, CRA may offer little or no greater or less resistance purely to erosion than CLAS of the same hardness. However, erosion-corrosion conditions are not addressed in this chapter as CLASs and lower Cr (<18%) containing stainless steels are susceptible to pitting, whereas higher CRAs may only suffer erosion. In addition, the present chapter only addresses internal corrosion threats imposed by production streams and excludes reference to both injected fluids and external corrosion.

In order to offer resistance to metal loss corrosion, a move to using CRAs would require a number of attributes including:

- strength comparable to CLAS alternatives;
- adequate resistance to EC and pitting corrosion in the intended service environments;
- compatibility with joining requirements, e.g. by welding, threading, flanging, etc.
- adequate availability of product form and quantity at an acceptable cost for each application.

No single alloy can offer these requirements for all duties. The established CRAs are therefore a sequence of Cr-containing alloys that starts with the leanest stainless steels containing around 13% Cr martensitic stainless steels and beyond with increasing amounts of Cr, Ni, Mo and other elements. The overall costs increase with additional alloying and manufacturing complexity. Beyond the Ni-based alloys, Ti-alloys provide further options that, though not yet widely used, may allow the development of hydrocarbons with source temperatures and pressures higher than current high pressure/high temperature (HPHT) sources referred to in Chapter 19. The metallurgy of Ti-alloys is distinctly different to that of stainless steels and Ni-alloys, therefore it requires caution.

3.2 Alloying Elements, Microstructures, and their Significance for Corrosion Performance

The combination and interaction of the alloying elements from different structures in the steel result in certain properties of the steel which in turn affect corrosion performance. This section deals with the interaction of metallurgical aspects including alloying elements, microstructures, and heat treatment and their combined influence on corrosion properties.

Typical types, categories, and grades of CRA are presented in Tables 3.1 and 3.2. It is evident that there is a wide variety of alloy choices, depending on the desired properties and, of course, the required corrosion resistance. This latter requirement is the major factor driving the selection of CRAs and is discussed in more detail later in this chapter.

It should be noted that 'high' strengths have to be balanced against increased sensitivities to EC. In particular, downhole CRAs that are susceptible to sulphide stress cracking (SSC) have limited tolerance of H_2S, so can only be used in sweet or reliably-mild H_2S-service. Thus, 'sweet' corrosion of CLAS can be combatted with the lower-cost/leaner CRAs, but more expensive/richer alloys are required for more severe H_2S-service conditions. There is a step change in H_2S tolerance 'beyond' duplex stainless steels (DSSs) as outlined in ISO 15156 [1].

3.2.1 Alloying Elements

Alloying elements present in CRAs fall basically into two groups.

- *Ferrite formers*: These limit or inhibit the formation of austenite, e.g. chromium.
- *Austenite formers*: These increase the level of austenite, e.g. nickel.

Cr and Ni are among the most important alloying elements here. All ferrite formers have a Cr equivalent and all austenite formers have a Ni equivalent as discussed later in this chapter.

Ferrite is important in avoiding hot cracking during cooling from hot temperatures encountered particularly in the welding of austenitic stainless steels. In such situations, 'constitution diagrams' are used to predict ferrite levels from the composition by comparing the effects of austenite and ferrite stabilising elements. Typically, the so-called Schaeffler and Delong diagrams are the original methods of predicting the phase balances in austenitic stainless steel welds and also used in predicting phases in CRAs (see Section 3.2.4) with the implications for each parameter on qualification and quality control covered in Section 3.2.4.1.

3.2.2 Improving Corrosion Resistance

Metal-loss corrosion resistance is improved by increased, but selective, additions of Cr, Mo, N, W, Nb, Ni, Co, and others. Carbon can be harmful through the 'depletion' of some additions as carbides, as described in Chapter 2. As the severity of service conditions increases, increased alloying content is required to resist both metal-loss corrosion and EC threats. Different types of EC threat that may affect the performance of CRAs are summarised and detailed in Chapter 6, including SSC, stress corrosion cracking (SCC) and galvanically induced hydrogen stress cracking (GHSC). ISO 15156-3 [1] lists primary and secondary concerns for generic CRAs in streams containing H_2S. Nevertheless, cracking that may be imposed by CP is excluded.

The metal-loss corrosion resistance of CRAs results from the formation of a tenacious, protective, Cr/Mo-rich passive oxide layer on exposed surfaces. A passive layer is only a few atoms thick (~10 nm), that forms spontaneously during dry atmospheric exposures and can remain protective in anoxic, hydrocarbon production fluids. In service, protective oxides may be (locally) removed by mechanical or chemical means; they therefore have to be capable of rapid self-repair. It should be noted that

Table 3.1 Summary of established CRAs and their nominal compositions.

Alloy group	Typical alloy and common name	Cr	Ni	Mo	Fe	C	Typical others	PREN range[a]
Martensitic SS	Families of 13Cr	11.0–14.0	0.5	—	Bal[b]	0.15		NA
Super martensitic SS	Alloyed 13Cr	11.5–17.5	4.5–6.5	0.7–2.5	Bal	0.03	0.18N	NA
Duplex SS	22%Cr	21.0–23.0	4.5–6.5	2.5–3.5	Bal	0.03	0.1N	35–40
	25%Cr	24.0–26.0	5.5–7.5	2.5–3.5	Bal	0.03	0.5 Cu, 0.3 N	37.5–40
	Super Duplex	24.0–26.0	6.0–8.0	2.5–4.0	Bal	0.03	0.25 N, 0.5W, 0.75 Cu	40–45
Austenitic SS	316L	17	12	2.2	Bal	<0.03		NA
	254 SMO	20	18	6	Bal	0.01	0.2N, 0.75Cu	NA
Austenitic Fe-based alloys	Alloy 28	26.0–28.0	29.5–32.5	3.0–4.0	Bal	0.03	1Cu, 2.5Mn, 1Si	NA
Austenitic Ni-based alloys	Alloy 825 (Incoloy)[c]	19.5–23.5	38.0–46.0	2.5–3.5	Bal	0.05	1Ti, 2Cu	NA
	Alloy G3	21.0–23.5	Bal	6.0–8.0	18.0–21.0	0.01	1W, 2Cu	NA
	Alloy 625 (Inconel)[c][1]	19.0–21.0	50 min	8.0–10.0	8.0–10.0	0.015	2.5Co, 4Nb, 0.2Ti	NA
	Alloy C276 (Hastelloy)[c][1]	14.5–16.5	Bal	15.0–17.0	4.0–7.0	0.02	0.2V, 4W	NA
Age-hardenable nickel alloys	718	19	52	3	Bal	0.05	5 Nb, 0.6 Al, 1Ti	NA
	725	21	57	8	Bal	0.01	3.5 Nb, 1.5Ti	NA
	945	21	50	3.5	Bal	0.01	3.5 Nb, 1.5Ti	NA
Titanium alloys	Ti-6Al-4V	—	—	—	—	—	6 Al, 4 V and Ti	NA
	Ti-6-2-4-6	—	—	6	—	—	6 Al, 2 Sn, 4 Zr and Ti	NA
	Beta C (38644)	6	—	4	—	—	3 Al, 8 V, 4 Zr and Ti	NA

Source: [1, 3, 4].

[a] As noted earlier in this chapter, while PREN = Cr + 3.3 (Mo + 0.5 W) + 16 N is an indicative of relative resistance to pitting corrosion threat, its applicability is primarily related to DSSs.

[b] Balance – NA; not applicable.

[c] A trade name.

Table 3.2 Typical range of yield strength for different categories of CRA.

Alloy group	Typical alloy and common name	Strengthening method	Typical yield strength, ksi (MPa)	Application			
				Downhole (API 5CT/ISO 11960 or API 5CRA/13680)	Pipeline (API 5LC)	Clad pipeline (API 5LD)	Well equipment/accessories API 6A
Martensitic SS	Families of 13Cr	Q/T[a] treatment	80–95 (552–655)[b]	✓	✓		✓
Super martensitic SS	Alloyed 13Cr	Q/T treatment	95–125 (655–862)	✓	✓		✓
Duplex SS	DSS	Solution annealed (SA)	65 (448)	✓	✓	✓	
		Cold worked (CW)/Cold hardened (CH)	125–140 (862–966)	✓			
	SDSS	SA	75 (517)	✓	✓	✓	
		CW/CH	125, 140 (862, 966)	✓			
Austenitic SS	Austenitic SS	SA	30 (207)		✓	✓	✓
Austenitic-Fe and Ni-based alloys		SA	45 (310)		✓	✓	✓
		CW/CH	110–140 (759, 966)	✓			✓
Age hardened (AH) nickel based		Age Hardened/precipitation hardened	110–160 (759–1104)				✓
Titanium alloys	α/β titanium alloys	Heat treatment	110–130 (759–897)	✓			✓
	β titanium alloys	Heat treatment	140–160 (966–1104)	✓			✓

[a] Q&T: Quench and Tempering heat treatment.
[b] Could be lower for pipeline applications.

the leaner alloys may suffer metal-loss corrosion, often highly localised in morphology, under water-wet atmospheric exposure if contaminants, particularly chlorides, are present.

In aqueous service, oxide repair is affected by oxygen obtained from the water-phase, If protection is 'lost', CRAs are prone to localised metal-loss (as crevice corrosion and/or pitting) and, in some circumstances, environmental cracking.

3.2.3 Pitting Resistance Equivalent Number (PREN)

Highly localised degradation threats including SCC usually initiates in corrosion pits and hence susceptibility to such threats can hint at the possibility of corrosion cracking. The pitting resistance of stainless steel is primarily governed by its composition, as referred to earlier. In this context, various indexes have been proposed to account for the relative benefits of alloying additions to improve the corrosion resistance of CRAs, although their use requires caution as they are indicative comparators only and do not necessarily account for resistance to all corrosion threat types.

CRAs with adequate metal-loss resistance for an intended duty have to be assessed also for aspects of EC. To help in the differentiation and ranking of such steels for a particular application, an equation called the pitting resistance equivalent number (PREN) has been developed. PREN is a theoretical way of comparing the pitting corrosion resistance of various types of stainless steels, based on their chemical compositions. However, it can neither be used to predict tendency to pitting nor cracking corrosion, nor whether a particular type will be suitable for all given application.

There are several variants of the equation with the one most widely referenced as Eq. (3.1):

$$PREN = \%Cr + 3.3(\%Mo + 0.5\%W) + 16\%N \qquad (3.1)$$

As a general rule of thumb, the higher the PREN, the better the resistance to pitting.

PREN is particularly applicable to duplex stainless steels (DSSs). As Ni is not included in the formula, its applicability to Fe-based and Ni-based austenitic stainless steels is questionable. Nevertheless, it can generally be used for ranking of such types of CRA.

Examples of typical PREN are given in Table 3.1. DSSs with PREN values above 32 are considered somewhat resistant to chloride-containing media. DSSs with PREN greater than 40 are classed as super duplex stainless steel (SDSS) and typically a minimum requirement for exposure to raw sea water, although this will have a temperature limit.

3.2.4 The Schaeffler Diagram and its Application

The Schaeffler diagram is an important tool for predicting the constitution of austenitic Cr-Ni steel and to represent the effect of the proportion of these elements and therefore the composition of the alloy on the structure obtained after rapid cooling from 1050 °C to room temperature. Compositions of the leaner CRAs can be illustrated on a Schaeffler diagram, and a typical version of this is shown in Figure 3.1. The position of some conventional steels are superimposed and depicted in Figure 3.1.

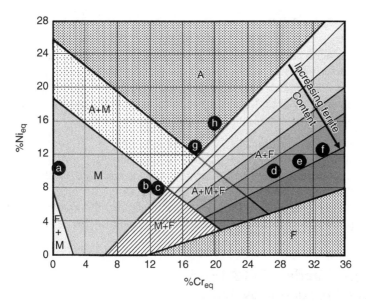

Figure 3.1 A typical Schaeffler diagram. (A: Austenite, M: Martensite, F: Ferrite) typical position of different steels: (a): Grade API 5CT L80 CLAS, (b): 13%Cr, (c): Alloyed 13%Cr, (d): 22Cr DSS, (e): 25Cr DSS, (f): 25Cr SDSS, (g): 304 SS, (h): 316SS.

It should be noted that ferritic areas (F) have mechanical and, potentially, toughness limitations and, therefore, respective alloys have low engineering use. Also, multi-phase regions produce complex microstructures that are potentially prone to multiple failure modes that make it difficult to determine service limits. These two areas are generally avoided when dealing with advantageous engineering alloys.

The Schaeffler diagram is especially suited to weld metals in order to predict the structure. However, it has been used by manufacturers to predict phase ratios in the development of new categories of CRA. Nevertheless, it should be borne in mind that due to inadequacy in determining the volume of a specific phase in multiphase zones and microstructural issues regarding the carbide formation process, its application needs to be used with great care.

While intended for a different purpose, the Cr equivalent, as explained below, is indicative of alloying constituents against metal-loss corrosion. Thus, stainless steels, with increasing metal-loss corrosion resistance, progress from left to right across the lower-part of the diagram; the trend then veers upward to show (more-costly) high-Ni stainless steels; Ni-alloys lie beyond the upper limit of the diagram so are not shown in Figure 3.1.

It suffices to note that engineered and practical compositions are now well defined after many years of commercial scrutiny. This does not exclude occasional new offerings enabled by improved manufacturing capabilities or specific service requirements.

3.2.4.1 Nickel and Chromium Equivalents

The Schaeffler diagram in Figure 3.1 shows the limits of austenite (A), ferrite (F), and martensite (M) phases in relation to the Cr and Ni equivalent. A 'nickel equivalent' (Ni_{eq}) is calculated for the austenite stabilising elements and a 'chromium equivalent' (Cr_{eq}) ferrite stabilising elements with typical formulae such as shown in Eqs (3.2, 3.3):

$$Cr \text{ equivalent} = Cr + Mo + 1.5Si + 0.5Nb \tag{3.2}$$

$$\text{Ni equivalent} = \text{Ni} + 30\text{C} + 0.5\text{Mn} \qquad (3.3)$$

with all concentrations being expressed in weight percentages. These are used as the axes for the diagrams, which show the compositional equivalent areas where the phases should be present.

It is of note that the Schaeffler diagram should be modified to provide a more accurate prediction of weld structure, bearing in mind the following points:

- The carbide formation process must be taken into consideration.
- Implementation of variable coefficients in Cr and Ni equivalents equations (the coefficients should depend on the concentration and mutual influence of alloying components, and on the carbide formation process in the weld).
- Incorporation of phase percentage lines for interphase zones (i.e. the zones which contain two or more phases, as performed by Schaeffler for austenite-ferrite zone).

3.3 Common Types/Grades of CRA Used in the Hydrocarbon Production Systems

There are many categories of CRAs and the list continues to grow as new products become available offering improved properties. Therefore, this section outlines only the most common alloys and categories.

CRAs are generally divided into groups or families of alloys that have common characteristics or microstructures. These are summarised in Table 3.1 and include:

- martensitic stainless steels (MSSs) and super MSSs (SMSSs)
- duplex stainless steels (DSSs) and super DSSs (SDSSs)
- austenitic stainless steels and super austenitic stainless steels
- iron-based alloys
- nickel-based alloys
- titanium-based alloys.

3.3.1 Nominal Compositions

The nominal chemical compositions of some of the most common alloys in each group above are presented in Table 3.1. A more comprehensive listing of alloys can be found elsewhere [1, 3, 4].

3.3.2 Mechanical Properties and Strengthening Methods

The application of the alloys in Table 3.1 to specific pieces of equipment for hydrocarbon production facilities depends largely on the strength of the alloys and the methods used to strengthen them. Some alloys are used in the solution annealed condition (SA) while others are strengthened by cold working (CW)/cold hardening (CH) at ambient temperatures and still other alloys can be heat-treated to obtain the desired strength levels. In the SA condition, invariably the yield strength does not exceed 80ksi (552 MPa).

Table 3.2 summarises the strengthening method, typical strengths, and relevant applications for some of the alloys listed in Table 3.1. Strengthening for downhole alloys used for well completions (tubulars) in particular is achieved by means of cold drawing, pilgering, ring expansion, and others. These may induce different mechanical and corrosion properties. Higher strengths can be achieved for 'critical' parts in accessories and very high-strength tubulars have been proposed and/or investigated. The strengths of downhole CRA tubulars are generally higher than generic variations used in gathering and processing facilities; this dictates metallurgical differences.

For CRA well completion, pipe, as production liners and tubing, is routinely selected first; thereby establishing the minimum metallurgy required for accessories.

3.3.3 Yield Strength

The yield strength of CW/CH pipe, including DSSs and richer alloys, is anisotropic, implying different properties in different directions, i.e. longitudinal against transverse. This is due to both residual stresses from CW causing the so-called Bauschinger effect as discussed in Section 3.3.3.1 and microstructural texturing. Anisotropy of yield is specific to the manufacturing process; particularly the final CH pipe-making process, including drawing, pilgering, ring expansion, etc., although other details of the manufacturing process also contribute through reduction ratios, finished sizes, etc.

Quantification of anisotropy is challenging as small specimens are compromised by relaxation during extraction with loss of residual stress, and full-thickness tests are generally impractical. Furthermore, full-thickness characterisation does not assure the behaviour of critical locations in connections that are locally machined. Anisotropy of yield can be accounted for by threaded connection testing of candidate products [5], however, there currently are no reliable methods to quantitatively predict the amount of anisotropy in casing and tubing. Tests establish a service envelope for the connection and, by association, potentially for the pipe. The latter is limited by the geometry of the connection test specimens so, for full qualification, the pipe may require additional testing of full-size collapse and burst specimens remote from connections.

3.3.3.1 The Bauschinger Effect

The stress-strain characteristics of any metal or alloy depend on its microstructure (atomic structure) and on its history, its manufacturing route, and heat treatment before use.

If a metal is initially stretched plastically in tension (loaded beyond the elastic limit) by a certain amount so that work-hardening occurs, and when unloaded, it will normally have a higher initial yield stress on subsequent loading in tension. This is conventionally referred to as 'work hardening' effect as schematically shown in Figure 3.2a which results in strengthening of an alloy by plastic deformation. In general, the effect of plastically working a metal is to increase its resistance to further working. Work-hardening inevitably results in a loss of ductility.

However, if the metal is subjected to compression after the initial stretching, it is found that the initial yield stress in compression may be substantially lower, often even lower than the initial yield stress before stretching. The lowering of the yield stress on reversal loading is referred to as the Bauschinger effect. This effect is an important parameter for CRA tubular and particularly when dealing with threaded connections. Both the Bauschinger and the work-hardening effects are schematically shown in Figure 3.2.

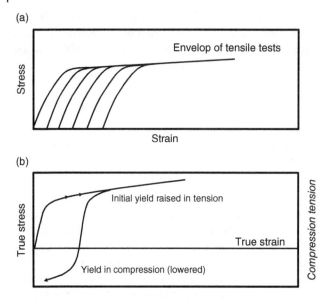

(a)

Stress

Envelop of tensile tests

Strain

(b)

True stress

Initial yield raised in tension

True strain

Yield in compression (lowered)

Compression tension

Figure 3.2 Schematic representation of (a) work-hardening and (b) the Bauschinger effect.

In plain terms, the Bauschinger effect refers to the differing yield point of metals and alloys when subject to cold working (low temperature mechanical treatment). Generally, the yield properties in tension and compression should be similar. However, due to the Bauschinger effect, in metals and alloys which have been pre tensioned, their yield point for further tension is raised whereas that for further compression is lowered. This is more critical if metals and alloys have directionality with anisotropic (non-homogeneous) properties in the longitudinal versus transverse directions. The ratio of yield strength in compression to tension is the so-called Bauschinger effect or ratio. This effect is particularly distinct for heavily CW/CH CRAs, since the majority of CRA tubular are strengthened to achieve a desired yield strength by mechanical treatment either cold drawing or pilgering, as outlined in Chapter 2.

The Bauschinger effect is not considered a major issue for SA materials (homogeneous) which may be used for expandable sand screens (ESS), pipeline or clad applications. The implication of this has not been fully addressed in the past. However, the Bauschinger effect for ESS manufactured from austenitic stainless steels (such as grades AISI 304 or 316 austenitic stainless steels), may not be critical: the pipe should start in fully SA (homogeneous conditions) and most of the material is not actually subject to much CW (mainly just opening up the slots on expansion). Nevertheless, if the component is subject to a few tension/compression cycles in ESS, then the Bauschinger effect may become an issue. Therefore, it is generally true that components which are to be used in tension should not be work-hardened by compressive loading at least in the direction of the tension.

3.3.3.2 Yield Stress Derating with Temperature
In general terms, tensile mechanical properties are affected by the prevailing exposure temperatures in a converse manner whereby increasing temperatures result in lowering yield as well as ultimate tensile strengths. This effect is particularly prominent for

Table 3.3 Typical yield strength derating of different steels.

Alloy	Derating of yield strength (%/100degC)
Carbon and low alloy steels	2–5
Families of 13%Cr	4–7
Duplex stainless steels	5–9
Austenitic Fe-based alloys	5–10
Austenitic Ni-based alloy	7–10

materials with isotropic microstructures such as mechanically worked CRAs and is less pronounced in SA or fully heat-treated alloys, although it still occurs to some degree. For CRA types/grades that achieve their strengths by heat treatment, structure derating of strength at elevated temperature is less affected than those that have been cold worked.

The yield strength derating effect has to be considered when designing wells and particularly in HPHT regimes. Table 3.3 can be taken as a guide for typical derating of yield strength of CRAs with temperature.

3.4 Important Metallurgical Aspects of CRAs

The thermal and to some extent mechanical processing history of CRAs can have a profound effect on the mechanical properties of these alloys and on their corrosion resistance. Some key factors of such processes and their respective consequences are summarised in this section, categorised according to the microstructure.

The discussion in this section focuses on wrought CRAs used for downhole components that are connected using threaded connections. However, many of these same alloys or similar alloys are used on the surface (or subsea for offshore applications) in facilities and pipeline/flowline applications, in which case, they are most often welded. The metallurgical aspects of welding these various alloys including the need to optimise their corrosion resistance are beyond the scope of this chapter. Suffice it to say, fabrication of some high types/grades of stainless steels, nickel alloys, and titanium alloys remains a significant challenge and it should not be assumed that all fabrication methods are now sufficiently established that there is no need for concern. In fact, many projects are delayed waiting on proper qualification of welding methods and procedures for CRAs.

3.4.1 Martensitic and Super Martensitic Stainless Steels (MSSs and SMSSs)

The MSSs and SMSSs are heat-treated by a quenching and tempering (Q&T) process to achieve the desired strength primarily for well applications through the formation of a martensitic phase. However, the process is complex, during which the alloys are prone to potential retention of two separate phases, including delta ferrite and retained austenite [6]. It should be noted that these phases do not normally form together. It has generally been found that these retained phases do not appreciably affect the metal loss corrosion resistance of MSSs and SMSSs, especially at low volume fractions. However, they can have a significant detrimental effect on mechanical properties and toughness with increasing volume fraction [7]. The transformation of retained austenite to untempered martensite could have negative effects on SSC resistance and tolerance to H_2S, although published evidence is limited.

Additionally, it has been found that increasing strength in the MSSs produces poor fracture toughness, especially where subsea applications may expose these alloys to temperatures close to freezing ($0\,°C$), such as at the mudline areas. The SMSSs have the advantage of sufficient Ni content to provide adequate toughness at low temperatures and for this reason are often selected for deepwater applications over the standard MSS types.

Increasing the Cr content to 15–17% provides improved performance in terms of metal loss corrosion and top temperature limits. However, these alloys have a tendency to form additional phases including delta ferrite and controlled heat treatment, and alloy compositions are paramount to minimise such deleterious parameters. It is generally recognised that while providing advantages, increasing alloying elements can lead to increasing metallurgical challenges and the leanest alloy, where applicable, is commonly preferred.

It should be noted that delta ferrite is highly textured during hot rolling, and that this usually significantly depresses its transverse properties. An example is that in relation to 17-4PH which is precipitation hardened alloy and not a tubular product but a valuable reference [8].

It is appropriate to emphasise that metallurgical expertise is needed to outline and make certain the implementation of basic methodology to produce these alloys.

3.4.2 Duplex and Super Duplex Stainless Steels (DSSs and SDSSs)

The DSSs and SDSSs are even more microstructurally sensitive than the martensitic steels since they are intentionally dual phase and each phase provides certain qualities and advantages over the other. DSSs and SDDSs derive their exceptional properties from the balance of phases between ferrite and austenite. Excess quantities of either ferrite or austenite cause the alloy to lose the beneficial properties of mechanical strength and corrosion resistance gained from a balance close to 50% for each phase [9–11]. The development and maintenance of this phase balance can become especially difficult for welds in DSSs. The generation of the sigma phase and other deleterious phases can impair the corrosion resistance and fracture toughness of these alloys [7]. Brittle failures of forgings and castings with a significant sigma phase have occurred in the petroleum industry, albeit infrequently. The development of the SDSSs has increased the application range of DSSs in severe corrosive service, although they do not significantly enhance resistance to cracking in the presence of H_2S compared to the DSSs. Once again, it is generally recognised that while providing advantages, increasing alloying elements can lead to increasing metallurgical challenges and the leanest alloy, where applicable, is commonly preferred.

It is also worth noting that high strength grades are normally obtained by cold working/cold hardening, and that this cold working rate is a key parameter affecting sensitivity to SCC in aqueous phases containing a combination of chloride and H_2S [12].

3.4.3 Austenitic Stainless Steels (SSs)

These categories of SS are not typically used for well applications except for control lines and some special equipment, such as sand screens. However, they are widely used for surface facilities and piping in corrosive field applications. The industry has made extensive use of AISI 316L lining for CRA-lined steel pipe for flowlines and pipelines and in some cases 904L has also been applied. However, while they have some resistant to SSC, their sensitivity to SCC is significantly enhanced in aqueous phases containing a combination of chloride and H_2S [13] as well as by cold working.

3.4.4 Austenitic Fe- and Ni-Based Alloys

The nickel-based alloys used in the hydrocarbon production industry are alloys that are strengthened either by: (i) cold working/cold hardening (CH); or (ii) precipitation hardened (PH) process that rely on the precipitation of gamma prime (γ') or gamma double prime (γ''). Subject to size requirements, tubes are supplied in CH conditions obtained by either cold pilgering or cold drawing for downhole applications. Moreover, Alloys 825 and 625 are widely used for internal cladding, particularly for manifolds (Alloy 625) and risers (Alloy 825), and flowlines (Alloy 825 and 625), although, certainly in the case of the latter – and 825 in general – great care needs to be exercised during exposure to hydrotest fluids as their pitting corrosion resistance in the presence of Cl^- and O_2 is limited.

Limits of application of these alloys are included in ISO 15156 in terms of H_2S tolerance and other operational parameters [1].

As with the SSs, PH nickel-based alloys have certain microstructures that can be prone to failure in certain petroleum production conditions. An example of this sensitivity that has resulted in several serious failures is the presence of acicular delta phase in Alloy 718 [14]. Alloy 718 is a nickel-based alloy that can be strengthened by the PH process. API 6A CRA [15] outlines the necessary measures to alleviate the microstructural anomalies that may render this alloy susceptible.

While Alloy 718 was cited as an example, all of the other PH nickel-based alloys used in the petroleum industry can form numerous deleterious phases such as sigma, chi, mu, etc. if they are not properly manufactured or are improperly fabricated, such as during welding [15]. However, these alloys have performed admirably with great success and the true limits of nickel-based alloys in petroleum environments have yet to be determined.

3.4.5 Titanium Alloys

Titanium alloys have a niche application within hydrocarbon production. However, bearing in mind that they have poor corrosion resistance when exposed to an environment containing trace fluoride ion, their downhole applications become limited if dealing with mud acid. In addition, while Ti alloys are somewhat immune to SCC in sweet or aerated brines, the standard types are sensitive in sour brines [11]. They are also prone to SCC in pure methanol, a widely used hydrate control chemical, however, small amounts of water significantly reduce the risk of SCC. Consequently, in hydrocarbon production only specific types/grades can be used.

For high strength downhole applications, micro-alloyed versions of the standard alpha/beta (α/β) grade Ti-6Al-4 V have been successively proposed. However, grade 24 containing Pd at times can be more costly, and the alternative Ru-containing version (grade 29), or the more alloyed Ti-6Al-2Sn-4Zr-6Mo (Ti 6246) were considered more favourably.

Purely beta (β) alloys were also considered, albeit by under-estimating the basic manufacturing difficulties associated with types/grades without any phase transformation (irreversible grain growth). Only one grade was finally registered as Grade 19, Ti-3Al-8 V-6Cr-4Zr-4Mo or Beta C [16]. It could be summed up that titanium alloys for downhole application still remain 'under consideration' [17].

The application of Ti alloys goes beyond wells, for example:

- Unalloyed Ti is the standard material for raw sea water applications, both in its welded or unwelded state, and has particular application for conditions where crevices cannot be avoided or CP cannot be introduced.
- α/β alloys such as Ti-6Al-4 V (TA6V or grade 5) are weldable, but less likely in practice, and caution is essential if considered. In particular, stress relieving under vacuum is not an easy process on welded parts, as they may suffer sustained load cracking (SLC), a purely metallurgical delayed rupture. These alloys are heat-treatable and can generate moderate to high strength and maintain high corrosion resistance.

Ti alloys require short ageing times to achieve strength, compared to the long ageing cycles required for strengthening nickel-based alloys. Analogous to steels, they also display a hardenability effect in thick sections. Ti alloys, however, have relatively low fracture toughness when compared to the nickel-based alloys. Due to their strain rate sensitivity, they often display poor Charpy toughness but under slow loading conditions can demonstrate adequate fracture toughness for most well applications. Some recent research has better defined the environmental limits of candidate Ti alloys for future use in oil and gas developments [15].

3.5 Limits of Application

The criteria for establishing environmental service limits or the window of application of CRAs include a combination of no or acceptable metal loss corrosion and no EC under the expected operating conditions.

Resistance to EC is subject to an alloy's composition and metallurgical condition, as routinely defined in manufacturing specifications. Environmental service limits are dependent on anoxic fluids in production streams. Primary environmental conditions for the occurrence of EC in CRAs include at least five key parameters of: the minimum in-situ aqueous pH, operating/maximum exposure temperature, maximum level of chloride in the water phase, maximum partial pressure of H_2S (or at higher pressures possibly maximum fugacity) and, finally, the presence of elemental sulphur,[1] all in related exposure conditions. These have been characterised by ISO 15156 Part 3 for the categories of CRA in Table 3.1.

It should be noted that oxygen, dissolved in water, is a potent cathodic reactant that restricts the service limits of CRAs. Therefore, an important distinction between the use of CRAs in hydrocarbon production and their application in, for example, sea water is the complete absence of oxygen in the former.

In the case of hydrotesting, discussed further in Chapter 7 (section 7.8), caution needs to be exercised. In such situations, the water may well be specified to be dissolved oxygen-free and treated with oxygen scavenger, however, often reality leads to something else especially where the corrosion expert is not directly on site – so untreated sea water may well be used for convenience, time, cost and due to the presumption 'well, it is a CRA, therefore it will not corrode'. That may well depend on how long the hydrotest water remains in a system and the attention to removal. The pitting found in the Alloy 825 riser internal cladding in the past was caused by poor management of the hydrotesting and wet parking. It should be highlighted that wet parking refers to leaving the riser on the sea bed prior to it being connected to the platform or floating production storage and offloading (FPSO) system or there being a delay in schedule for connection thus either leaving the hyrotest water in the riser for longer than prescribed with the potential for the chemical treatment package present becoming increasingly less effective (being consumed and/or degraded by chemical/biological reastion) – falls below optimum concentration – and/or ingress of a raw sea water during the hold up. Such related corrosion damages are discovered several years after start-up when a first smart pig is run or by pure chance. The consequences of such damages are costly and worrying.

CRAs used in hydrocarbon production are used at higher chloride contents and higher temperatures then ever possible when oxygen is present, such as may occur in surface equipment and during water injection. The presence of oxygen in surface equipment, even in the parts per billion range, generally limits the use of CRAs to <100 °C.

3.6 Selection Criteria

Materials selection criteria are dealt with in Chapter 14. However, as previously indicated, suitability for joining is a discriminator for CRA applications for surface facilities and flowlines. Simplistically, welding considerations differentiate downhole CRA applications, based largely on pipe with threaded connections, from welded gathering systems, processing facilities, and pipelines.

Downhole CRA pipe is not normally welded. Downhole equipment, wellheads and Christmas-trees may all be welded but, when required, welding is done in workshops that allow post-weld heat treatment (PWHT) when necessary. This contrasts with the norm of 'as welded' construction for gathering pipelines and primary pipework in processing facilities. The result is that, although similar generic materials may be used, detailed metallurgies differ. Also there is a need to check compatibility with exposure to external CP if a line is buried, even though it will likely be coated and insulated.

3.6.1 Selection Criteria Check List

In consideration of CRAs for a particular application, many aspects need to be reviewed and taken on board. Apart from economy and appropriateness as outlined in Chapter 14, in principle, two key elements are dominant: qualification and quality control:

- *Qualification*: Qualification tests can be carried out on candidate materials and production routing to qualify for a particular duty. This is usually carried out before placing an order, on samples representative of the expected worst case in production. A qualification test is specific to an application and will only be carried out in the absence of existing data on similar grades.

- *Quality control*: Quality control tests can be carried out during the manufacturing or purchase from the open market to verify the properties if required.

Purchase of CRAs is normally done in accordance with certification following a particular international specification. Table 3.4 outlines some key typical criteria when

Table 3.4 Selection (check list) criteria for aspects of corrosion and mechanical properties of CRAs (only for the production streams).

No	Performance indicator	Qualification	Quality control	Notes
1	Mechanical properties (yield strength and toughness)	As per ASTM A370 in both directions	Subject to discussion	Functional requirement
2	Metal loss corrosion resistance (by addition of Cr, Mo, etc.) – refer to Items 4 and 5	Documented field history or corrosion testing may be historical and/or application specific	Appropriate specification and manufacturing oversight (mill certificates)	
3	Environmental cracking resistance in accordance with ISO 15156-3	According to ISO 15156-3 and/or NACE TM0177	Slow strain rate tensile esting (SSRT) according to NACE TM0198	
4	PREN value	Only applicable to Duplex Stainless Steels and in particular injection systems, although an indication of potential corrosion in chloride containing media		
5	Indication of Cr equivalent content and phases for DSS	Schaeffler diagram		
6	Localised metal loss corrosion (crevice corrosion and pitting tests)	Documented field history or corrosion testing may be historical and/or application specific	Options: ASTM G48 although not relevant for the application and crevice	
7	Hardness profile	As per ISO 6507-1		For information
8	Yield strength derating	As per ASTM A370 at max operating temperature	—	
9	Anisotropy (Bauschinger effect, cold hardened for downhole only)	Option: enhanced connection testing	Manufacturing procedure specification (MPS)	
10	Joining	Tubular: connection testing for tubular Pipelines: weld performance and consumable Flange testing	Subject to discussion	

considering CRAs for duty in a particular production stream. Similar criteria are used for secondary environments, such as acidizing, completion/packer fluids, etc.

3.6.2 Application of CRAs

A brief overview of relevant information in terms of use for CRAs in included in Chapter 14. This is by no means exhaustive and only makes a reference to other more comprehensive sources. It is apparent that CRAs can be used in all elements of the production facilities where system corrosivity exceeds the tolerance of CLASs with or without corrosion mitigation methods. This is as outlined in Table 3.2, including wells, wellheads and templates, manifolds, subsea/surface facilities, flowlines, riser systems, gathering stations, vessels, and pipelines.

Two additional corrosion threats that should be noted are:

i) CP is an important consideration as a potential cause of GHSC of CRAs, although not considered in the ISO 15156 [1]. This is particularly applicable to flowlines and pipelines, and structures where CP is used to mitigate external corrosion but not common for downhole applications.
ii) Microbial viability, hence potential for microbiologically induced corrosion (MIC), is also a consideration for the leaner CRAs and can affect DSSs, particularly where there is intermittent use/exposure and dead legs.

3.6.3 Notable Points to Consider for Well Completion

The application of CRAs for well completion is subject to many practical parameters, notable examples of which are summarised in Figure 3.3. Some of these parameters are beyond the intent of the current chapter, although a few are included as described earlier or included in other sections or chapters. It is worth noting that while Figure 3.3 primarily focuses on downhole application, most of the parameters are equally applicable to other application scenarios.

3.7 Future Demands and Requirements

As the petroleum industry continues its quest for oil and gas, pushing to deeper horizons and harsher conditions, that are accompanied by higher pressures and temperatures, outlined in Chapter 16, a way forward may have to be through the increased use of CRAs. Despite this, much remains to be learned about these alloys and their limits of application. There are current limitations on the size and strength of CW/CH CRAs for some HPHT applications. Cost also plays an important role and moving to cost effective categories of CRA is desirable, bearing in mind production capacity against demand.

Lower weight alloys such as Ti alloys, if proved appropriate, can provide savings in whole life cost, not only for reduced weight on the platforms compared to steels and nickel alloys but also reduced hook loads when running the tubing. Therefore, the future in oil and gas production will require innovative solutions to complete and produce both deep HPHT wells and also high rate gas wells offshore. These solutions will include the use of CRAs.

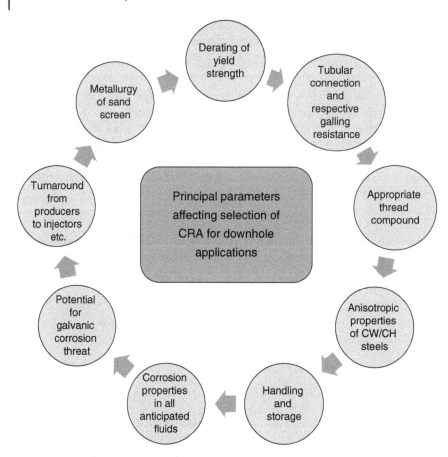

Figure 3.3 Principal parameters affecting the selection of CRA, in particular for downhole application.

3.8 Summary

CRAs are a group of materials used in hydrocarbon production with superior metal loss corrosion resistance. However, their manufacturing process and production route are critical and subject to metallurgical scrutiny in terms of correct compositions, heat, and mechanical treatments to allow production of a fit for service alloy. In terms of whole life costing, they may offer suitability for particular applications. Many challenges face their deployment, including microstructural homogeneity, anisotropy, mechanical performance, and potential corrosion threats not directly related to metal loss. Corrosion threats which are potentially likely for CRAs include all aspects of corrosion cracking, pitting under extreme conditions of high temperature and chlorides, and potential susceptibility when subject to CP and MIC.

CRAs are generally sub-categorised by their microstructural features; some are capable of offering high strength while others may have low yield strength and are unsuitable for all applications.

When considering CRAs for a particular application, many aspects need to be reviewed and taken on board. These are briefly outlined in the present chapter. Though discussed, the list is by no means exhaustive. Apart from economy and appropriateness, in principal, two key elements are dominant – qualification and quality control.

Note

1 Elemental sulphur is a corrosive cathodic reactant that, as a solid or liquid, has a high chemical activity. It may typically accompany H_2S when the gas-phase H_2S is above ~10 mol%.

References

1 ISO, Petroleum and gas industries – materials for use in H_2S containing environments in oil and gas production, ISO 15156, 2015.
2 Shreir, L.L., Jarman, R.A., and Burstein, G.T. (eds.) (1994). *Corrosion*, 3e. Oxford: Butterworth-Heinemann.
3 Craig, B. (2014). *Oilfield Metallurgy and Corrosion*, 4e. Denver, CO: MetCorr.
4 PISO, Petroleum and natural gas industries – corrosion-resistant alloy seamless tubes for use as casing, tubing and coupling stock – technical delivery conditions. ISO 13680, 2010.
5 API, Procedures for testing casing and tubing connections. API RP 5C5, 2017.
6 Kimura, M., Miyata, Y., Toyooka, T., and Kitahaba, Y. (2001). Effect of retained austenite on corrosion performance for modified 13% Cr steel pipe. *Corrosion* 57: 433–439.
7 Bilmes, P.D., Solari, M., and Llorente, C.L. (2001). Characteristics and effects of austenite resulting from tempering of 13Cr-NiMo martensitic steel welds. *Materials Characterization* 46: 285–296.
8 T Cassagne, M Bonis, C Duret, and JL Crolet, Limitation of 17-4PH metallurgical, metallurgical and corrosion aspects. NACE Corrosion Conference, Paper No. 03102, 2003.
9 Floreen, S. and Hayden, H.W. (1968). The influence of austenite and ferrite on the mechanical properties of two-phase stainless steels having microduplex structures. *Trans ASM* 61: 489–499.
10 JL Crolet, S Corbineau, and CP Perrollet, Sigmatization of duplex forgings: a case history. NACE Annual Corrosion Conference, Paper No. 15, 1997.
11 T Cassagne, and F Busschaert, A review on hydrogen embrittlement of duplex stainless steels. NACE Annual Corrosion Conference, Paper No. 05098, 2005.
12 JL Crolet, and MR Bonis, Evaluation of resistance of some highly alloyed stainless steels to stress corrosion cracking in hot chloride solutions under high pressures of CO_2 and H_2S. NACE Annual Corrosion Conference, Paper No. 232, 1985.
13 T Cassagne, G Moulié, and CD Correx, Limits of use of low alloy and stainless steels in upstream sour environments. NACE Annual Corrosion Conference, Paper No. 09079. 2009.
14 S Mannan, E Hibner, and B. Puckett, Physical metallurgy of alloys 718, 725, 725HS and 925 for service in aggressive corrosive environments. NACE Annual Corrosion Conference, Paper No. 03126, NACE, 2003.

15 API, Age-hardened nickel-based alloys for oil and gas drilling and production equipment. API Standard 6ACRA, 2015.

16 RD Kane, S Srinivasan, K Ming Yap, and B Craig, A comprehensive study of titanium alloys for high pressure high temperature (HPHT) wells. NACE Annual Corrosion Conference, Paper No. C2015–5512, 2015.

17 B Hargrave, M Gonzalez, K Maskos, et al. Titanium alloy tubing for HPHT OCTG applications. NACE Corrosion Conference, Paper No. 10318, 2010.

Bibliography

Specifications

API, CRA Clad or lined steel pipe. API Specification 5LD, 2015.
API, Specification for CRA pipeline. API Specification 5LC 2015.

Further Reading

Craig, B.D. and Smith, L. (2011). *Corrosion Resistant Alloys in the Oil and Gas Industry: Selection Guidelines Update, 3.* Toronto: The Nickel Institute, Publication No. 10073.

ISO, Petroleum and natural gas industries – steel pipe used as casing or tubing for wells. ISO 11960, 2010.

NACE, Petroleum and natural gas industries – materials for use in H_2S containing environments in oil and gas production orrosion-resistant alloy seamless tubes for use as casing, tubing and coupling stock – technical delivery conditions. NACE MR0175/ISO 15156, 2001.

4

Water Chemistry

Hydrocarbon production is associated with liquid water from a variety of sources. Water is either produced with hydrocarbons or is injected back into a reservoir to enhance recovery or for environmental reasons. Dry hydrocarbon by itself is not corrosive, it is the presence of water, or rather various species associated with the water, that causes corrosion in upstream production systems. Such species include dissolved acid gases, such as CO_2 and H_2S, naturally occurring organic acids, and acids introduced through production chemicals. Furthermore, the salinity and nature of the dissolved ions influence the likelihood and rate of corrosion.

At the design stage of a project it is essential to make judgements about the likelihood of corrosion in the various streams, as discussed in Chapters 5–7. Among other things, judgements have to be made about the likely composition of the various process fluids and how these change with time and evolving process conditions. Obtaining a truly representative water sample at the design stage applies equally to all the other reservoir fluids and gases, a significant subject on its own and potentially beyond the scope of the present chapter and publication. It should be highlighted that this is where the expertise of the reservoir engineers comes into play in determining how representative over the whole life of the field, the samples from exploration/test wells, etc. are, and then modelling how they will change over time as a reservoir becomes depleted and is subject to water and/or gas injection. Furthermore, all the meaningful and most reliable analysis of water occurs during the production phase – ensuring what is actually experienced does not vary significantly from that based on exploration wells and, if it does, to bring it under control so the as-built design remains fit-for-purpose.

Therefore, accurately defining water chemistry is a vital element in predicting and managing corrosion and materials performance during hydrocarbon production. The geographic location of the field, the reservoir structure, and the strata from which the hydrocarbon is produced, the temperature and pressure and the type and nature of the hydrocarbon all significantly affect the physical and chemical properties of produced water and its corrosivity. The volume and nature of these waters can also vary throughout the lifetime of a reservoir.

This chapter briefly describes the sources of different waters encountered in hydrocarbon production, the key characteristics affecting the corrosion of metallic materials, and the analytical methods used to determine water chemistry. Once

Corrosion and Materials in Hydrocarbon Production: A Compendium of Operational and Engineering Aspects, First Edition. Bijan Kermani and Don Harrop.
© 2019 John Wiley & Sons Ltd. This Work is a co-publication between John Wiley & Sons Ltd and ASME Press.

again the chapter is by no means exhaustive and only serves as an indication of important parameters. A more detailed discussion of the subject is undoubtedly beyond the scope of this publication for which reference is made to other key publications.

4.1 Sources of Water

Water is invariably produced with hydrocarbon products. In the early stages of a development, produced water volumes relative to the hydrocarbons produced tend to be low. However, with time, these water volumes tend to increase as attempts are made to extract increasingly more oil or gas from the reservoir. In mature developments, more than 98 barrels of produced water for every two barrels of produced oil may be encountered before the field becomes uneconomical to produce further.

Water may be produced from free water in zones that underlie the hydrocarbons or in the same zone as the oil and gas [1–7]. This originates either from waters percolating through surface or subsea rocks, building up into substantial aquifers over geological times, or as water trapped in sediments as they form into rocks. These waters can conveniently be referred to as formation waters, or, due to them commonly containing significant amounts of dissolved salts, formation brines.

Water is often injected into a reservoir to overcome the drop in pressure as hydrocarbons are produced, helping to maintain the oil production rates, as outlined in Chapter 7. This injected water can migrate with time towards production wells and then be co-produced in different proportions with formation brine. The produced water itself may be a convenient source for such injection water, or alternatively sea water or river waters may be used. These waters and indeed other formation brines experience temperature and pressure changes, and encounter different rock types and hydrocarbons over time, which changes the nature of the waters as they dissolve or precipitate different species. Further, pressure and temperature changes, experienced as fluids are produced at the surface, can reduce the ability of the hydrocarbons to hold water, resulting in condensed water being produced. Consequently, there can be extremely large variations in the chemistry of the water produced from different reservoirs or from the same reservoirs with time.

Types of water encountered in hydrocarbon production therefore fall broadly into three categories [7]:

- *Reservoir or formation waters:* this type of water contains many dissolved minerals and gases. Depending on the partial pressure of the acid gases, such as CO_2 or H_2S, its in-situ pH is normally in the range 4–7.
- *Injected water:* this is supplied from the surface and its dissolved minerals depend upon the source. If it is sea water, typically its pH would be in the range 7–8 and with many dissolved minerals. River water is often much fresher (less dissolved minerals and lower salinity) and often in the pH range of 6.5 –8. If taken from the water produced with hydrocarbons, the pH would be typically <7 as some dissolved acid gas would still be present.
- *Condensed water:* this is produced as the pressure declines when hydrocarbons are produced. In the case of gas wells, this type of water can make up the majority of the

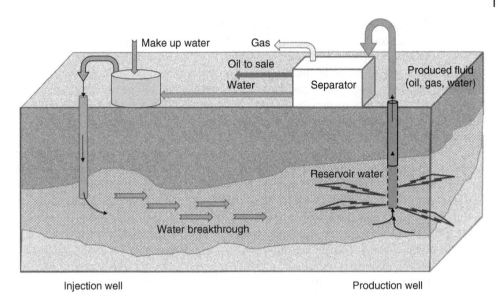

Figure 4.1 Schematic representation of sources of water in hydrocarbon production.

water produced. For oil wells, this type of water may amount to only a small fraction of the water volumes produced. This water tends to have low levels of dissolved minerals, and depending on the partial pressure of acid gases, its in-situ pH is normally between 3.0 and 4.5.

Figure 4.1 represents schematically the sources of water in hydrocarbon production.

Produced water chemistry is therefore highly dependent on the nature of the reservoir (e.g. hydrocarbon type, rock type, temperature, pressure, geological history) and the reservoir management strategy (e.g. well selection, water injection, produced water re-injection, production rates). To add a further complication, to produce oil and gas and to inject fluids safely and efficiently, different chemicals such as scale inhibitors and others are added. These can also alter the water chemistry, affecting corrosivity and materials selection and management.

Water chemistry is discussed in more detail in Section 4.2, and how this can impact corrosion.

4.2 Water Chemistry

The chemistry of the water is a critical input to corrosion modelling (see Chapter 5 for corrosion predictions for carbon and low alloy steels). It is also required to predict the risk of inorganic scale deposition. Therefore, an understanding of water chemistry is required during the project stages of a development to ensure the correct design and also during the production phase to confirm the initial assumptions and to ensure that the management strategies for corrosion and scale are effective and remain appropriate.

Water chemistry is a term used to describe the nature and levels of the dissolved minerals, the dissolved gases, the organic acids and the water pH. To determine the water chemistry, it is required that a representative water sample is taken and analysed. It is the chemistry of the water in-situ that is of real interest, i.e. at the pressure, temperature, and flow conditions at the region of interest. (It should be noted that unfortunately it is not possible to sample from all locations of high risk. Furthermore, most analyses can only routinely be carried out on atmospheric samples, and over time, the water chemistry of the sample will change.)

The main effects on in-situ water chemistry are from temperature and pressure changes which result in physical and chemical changes. A reduction in pressure can create a separate gas phase which reduces the amounts of gases dissolved in the water. For example, CO_2 loss will increase the pH, changing the risk of corrosion and scale. A reduction in temperature reduces the solubility of calcium carbonate in water, increasing the risk of scale, which consequently could reduce the remaining levels of the scaling ions dissolved in the water. Furthermore, many species are divided between oil, water, and gas phases in relative amounts, depending upon pressure and temperature. To understand these effects usually requires some additional information on the nature of the hydrocarbons and the producing conditions.

The physical and chemical changes to the water may take some time to reach equilibrium, and hence, once sampled, the water chemistry may also change. For instance, scale may take some time to form or over longer timescales, bacterial activity may result in loss of organic acids. Following the correct procedures for water sampling and analysis can minimise the water chemistry changes after sampling (see Section 4.4). However, the limitations on the information that can be obtained on water chemistry from specific samples and how this relates to the true nature of the water chemistry *in situ* usually require more than a simple sampling and analysis programme of a produced water stream. As a rule, it is advisable to determine brine pH and alkalinity at the sample point at the time of sampling, and repeat the analysis at the time the sample is analysed properly, because these are critical to corrosion and scaling risks and can change rather quickly.

Table 4.1 shows a typical water analysis and information sheet that could form the basis for determining the water chemistry for corrosion and scaling risks.

It should be noted that analysing for these species is usually conducted to determine scale risk. However, the same information is vital for establishing and managing the corrosion risks.

4.3 Other Impacts on Corrosivity

In addition to the water chemistry itself, there are several other secondary impacts on corrosivity resulting from chemical and physical changes to the chemistry of oilfield waters. In particular, any solids that form in a system can lead to under-deposit corrosion. Common examples of such solids include scale, bacteria, or iron sulphide. Any corrosion assessment or management strategy therefore needs to consider the risks of such solids. A detailed discussion of each of these is beyond the scope of this book, but each is touched on for completeness.

Table 4.1 Information on a typical water analysis sheet.

Water analysis report

Operator name, location and details

		Analysis of sample at room temperature		
Summary		**Anions**		**Cations**
Sampling date		Chloride	$mg\,l^{-1}$	Sodium	$mg\,l^{-1}$
Temperature at sampling point		Bromide	$(meq\,l^{-1})$	Potassium	$(meq\,l^{-1})$
Pressure at sampling point		Iodide			
Analysis Date and time		Bicarbonate		Magnesium	
Analyst		Carbonate		Calcium	
TDS	$mg\,l^{-1}$	Sulphate		Strontium	
Density	$g\,cm^{-3}$	Phosphate		Barium	
Anion/Cation Ratio		Borate		Iron	
Carbon dioxide in gas	ppm	Silicate		Aluminium	
Alkalinity at sampling point		Nitrate		Chromium	
BHP	psi/MPa	Formic acid		Copper	
BHT	°C/°F	Acetic acid		Lead	
Depth of sample	m/ft	Propionic acid		Manganese	
WHP	psi/MPa	Butyric acid		Nickel	
WHT	°C/°F	pH at time of sampling	–	Mercury	
CO_2	mol%	pH at time of analysis	–		
H_2S	mol%	pH used in calculation	–		

Notes: BHP; Bottomhole Pressure; BHT; Bottomhole Temperature; TDS; Total Dissolved Solid; WHP; Wellhead Pressure; WHT; Wellhead Temperature.

4.3.1 Mineral Scale

The anionic and cationic species shown in Table 4.2 can be used to indicate the risk of scaling. If scale is formed, not only is there an increased risk of corrosion under the scale, but the formation of scale means that samples taken downstream of where the scaling takes place will be depleted in scaling ions and, as such, the analysis may be accurate, but the results misleading or even irrelevant. Evidence from changes in scaling ion concentrations from one location to another may be used to establish whether scale deposition or precipitation is taking place and if so, where. Sulphate scales can

Table 4.2 Common analytical techniques for brine components.

Analysing for	Method of analysis
pH, bicarbonate, carbonate, alkalinity, hydroxide	Potentiometric titration at sample site or in laboratory
Chloride	Titration in laboratory
Sulphate	Ion selective electrode, inductively coupled plasma (ICP) atomic emission spectroscopy or turbidimetric in laboratory
Phosphate	ICP spectroscopy or visible spectroscopy in laboratory
Borate, silicate, sodium, magnesium, calcium, strontium, barium, iron, potassium, aluminium, chromium, copper, zinc, lead, manganese, nickel	ICP spectroscopy in laboratory
Organic acids	High performance liquid chromatography or ion exclusion chromatography in laboratory
Conductivity	On site or in laboratory
Carbon dioxide	On site (gas detector tubes)
Hydrogen sulphide	On site (gas detector tubes)

concentrate radioactive ions within deposited solids that are normally dissolved in the brines at very low amounts. The presence of this NORM (naturally occurring radioactive material) can be detected using a Geiger counter and evidence of this would suggest that scaling ions have been lost from the water.

4.3.2 Bacterial Analyses

Bacteria in oilfield systems may be either planktonic (in the bulk fluid) or sessile (attached to a surface), which determines both how samples are taken and analysed. Bacteria may be natural or introduced through human activities through, for example, water injection or via drilling muds. The brines and hydrocarbons provide the nutrients and food that allow bacteria to thrive. Their rate of growth is dependent upon the levels of nutrients, the temperature, and the salinity of the brine. High salinities tend to inhibit or slow bacterial growth and hence knowledge of the water chemistry also influences the bacterial management strategy. Their metabolic products, such as H_2S, organic acids, CO_2, or S, are the causes of increased system corrosivity, as outlined in Chapter 11, but also alter the measured water chemistry of samples downstream of any bacterial activity. The bacterial detritus can form a tough coating around bacterial colonies that protect them from flow, allow corrosion cells to be set up beneath the solids and reduce the effectiveness of any corrosion management techniques. This subject is covered further in Chapter 11.

4.3.3 Iron Sulphide

Iron sulphide is frequently a by-product of corrosion in sour systems where fluids contain H_2S and is common in water injection systems where bacterial-related corrosion is prevalent. This finely divided black solid is pyrophoric when dry and hence requires careful disposal.

It is also worth commenting here on other sulphur-related species present in produced fluids, such as mercaptans, thiols, and polysulphides, that influence corrosivity. These are an example of species that are distributed between oil, gas, water, and solid phases and may act as either corrosion inhibitors or enhancers. These materials are sensitive to changes in pH and oxidation and can produce elemental sulphur [8]. This area of water chemistry and the effect of various sulphur species on corrosion is complex and to some extent dealt with in Chapters 5–7.

4.3.4 Other Chemicals

Chemicals added to production and injection streams can alter the water chemistry through the nature of the chemicals. For instance, additional nitrogen and sulphur species from corrosion and scale inhibitors can add to the true levels of these species in water. It should be also pointed out that some chemicals can be corrosive themselves (e.g. some scale inhibitors are extremely acidic).

Corrosion of aluminium by mercury is a potential issue in gas processing but since mercury does not influence the corrosion of carbon and alloy steels, this is not discussed further here. However, mercury analysis in produced water, oil, and gas streams is recommended to allow precision and the necessary precautionary measures on the choice of materials.

4.4 Water Sampling Locations and Analysis Techniques

Table 4.2 summarises the typical water properties that are analysed, the common analytical methods and the location where they are normally carried out. Clearly it is not necessary to analyse for all of these species in every water sample taken, but as with any sampling programme, the reason for taking the sample will dictate the information required.

Some brines contain high levels of organic acid anions that can interfere with bicarbonate alkalinity titrations. Furthermore, the addition of other chemicals to the system, such as neutralising amines, hydrogen sulphide scavengers, or water clarifiers, can all affect colorimetric end points. Careful titration procedures and further analyses, such as amine determinations, may be necessary to account for such species. This is discussed further below.

4.4.1 Sampling

Correct sampling techniques underpin the generation of correct water chemistry. Where to sample, the volume of the sample, how samples are transported to the analytical laboratory, and the type of analysis are all critical considerations. This is also the case for information required from hydrocarbons, gas, or solids, this section provides a brief overview of water sampling. References to a number of publications [9–12] on the subject are suggested for more detailed information.

It is the water chemistry in-situ that needs to be understood. However, it is not possible to sample at every location where a corrosion risk exists, particularly in wells and production or injection lines. Even with careful selection of sampling locations, it is important to use the samples taken to understand the risks elsewhere in the system. The temperature and pressure changes, the other hydrocarbons phases which waters encounter, and the flow conditions all affect the in-situ water chemistry.

A sample should aim to be representative of the fluids at the sample location [14], and aim to remain representative until the time the analysis is carried out. In a multiphase system (where more than one of the phases oil, water, and gas co-exist), it is very unlikely that a multiphase sample will contain the exact ratios of each phase at the sampling point, even if these remain constant with time. It is always preferable therefore to take single phase samples from single phase lines if possible. However, single phase water samples will have experienced significant changes from waters further upstream. Temperature and pressure changes may cause mineral salts to precipitate and be lost from the water, or gases to be released and reduce the amounts of acid gases dissolved in the water. To complicate matters, it may take some time for a water at a particular location to reach equilibrium with its surroundings. It may take some time for the dissolved gases to be lost (e.g. it can take a fizzy drink some time to go flat after opening) and waters can remain oversaturated for some time before precipitation of scale.

It is often not possible to take single phase water samples, particularly if sampling upstream of production separators or from downhole sample tools. It is therefore, common practice to have to separate water from a multiphase sample where water is present with either oil or gas. The separation process, especially if the hydrocarbons are heavy black oil or contain species such as asphaltenes that can emulsify oil and water, may take some time, but is usually speeded up by temperature or addition of chemical demulsifiers. This may preclude analyses at the sample point, and if it becomes necessary, the details should be noted and supplied with the sample. Furthermore, if the volume of separated water is small, a full analysis may not be possible, leading to the operator having to prioritise the data required. The process of this separation can also alter the water chemistry if separation is carried out at a different pressure and temperature than at the sample location. It is unlikely that the sample will contain the exact ratio of the different phases at the location, so to obtain accurate data on species that partition between the different phases, further sampling and analysis may be required to specifically address this.

Whether sampled single phase or multiphase, water may be contaminated by drilling, the completion or stimulation fluids used during the well operations [9–12].[1] Hence it is always necessary to request information on other operations taking place and any chemicals added upstream of the sample point, or to cease such operations and turn off chemicals before sampling.

Sampling from high pressure systems requires particular attention, not least from the health, safety, and environment (HS&E) considerations. Sampling equipment must be rated to the appropriate pressure and samples collected via on-stream collectors. Alternatively, samples may be taken either into a piston vessel to control any pressure loss until analysis in a laboratory is possible, or flashed across the sampling valve into an evacuated vessel to ensure that any condensed liquids are captured. The physical processes experienced by the fluids will affect the water chemistry subsequently determined and so sample collection, sample separation, and sub-sampling

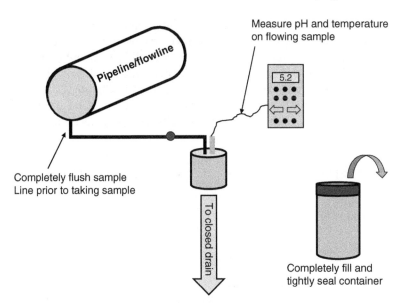

Figure 4.2 Typical sampling procedure at low pressure.

conditions must be known and noted to ensure that any changes can be taken into account for affected species.

In low pressure systems with high brine content, sampling is relatively simple and it is this type of sampling that is most common in determining water chemistry. Figure 4.2 illustrates the important points.

Samples can be taken at atmospheric pressure after the sampling lines have been flushed clean and any oil/water separation carried out in stoppered bottles. The sample bottle should be clearly labelled with date, time, sample point, temperature, and pressure at the sample point, name of person sampling, any known hazard (bio, radio, flammable, etc.) and any additives used (quantity and type) in stabilising the sample.

Ideally, samples should be taken upstream of any likely precipitation of solids, in order to avoid sampling a stream that has become depleted in some ions. Analysis from multiple sample locations may, however, be used to confirm whether scale deposition or iron sulphide precipitation is taking place. The picture for iron sulphide may be complicated by variable composition and solubility with time and morphological type [13]. Oxygen ingress should be minimised as any oxidation can alter the solubility of some species, particularly those containing iron, and affect bacteria viability.

4.4.2 Interpretation of Results

Obvious analytical errors or anomalies can be checked, for example, by calculating the anion/cation ratio and ensuring that this ratio is close to neutrality, recognising that some minor species may not have been analysed.

Produced water may contain increasing amounts of injected water with time. The scaling risk alters significantly with the change in proportions of these two types of water and hence it is important to know this ratio when interpreting scaling tendency results and residual ion levels. Ions that have sufficiently different concentrations in the two types of water can be used as tags to monitor the proportions of the two waters.

The corrosivity of a water and its scaling potential are critically dependent upon the pH of the water. A low pH (acidic) tends to increase corrosivity and a high pH (alkaline) tends to increase the calcium carbonate scaling risk, as outlined in Chapter 5. The in-situ pH is influenced by three controlling buffer systems [15–20]: CO_2/bicarbonate/carbonate; H_2S; and organic acids, with the former generally being the most important. The following chemical equilibria apply[2]:

$$CO_2(g) \Leftrightarrow CO_2(aq) + H_2O \Leftrightarrow H_2CO_3(aq) \Leftrightarrow H^+ + HCO_3^- \Leftrightarrow H^+ + CO_3^{2-}$$

$$H_2S(g) \Leftrightarrow H^+(aq) + HS^-(aq)$$

$$CH_3COOH \Leftrightarrow H^+(aq) + CH_3COO^-(aq)$$

The common factor is that each of these species (CO_2, H_2S, and organic acids) dissociate in waters to form H^+ ions. The concentration of H^+ ions defines the pH (the higher the concentration, the lower the pH). Therefore, any effects that move these equilibria towards the formation of more H^+ ions will reduce the pH. Higher temperatures reduce the solubility of CO_2 and H_2S in water while higher pressure increases the solubility. Hence, when the reservoir fluids are produced at the surface, these species will separate differently between the oil, water, and gas phases. As pressure is reduced, a gas phase is produced, CO_2 partitions from the water phase and the pH increases.

Any dissolved calcium ions in the water can react with the CO_3^{2-} ions moving the equilibria in the opposite direction, continuing to consume bicarbonate and hence CO_2 until the solubility limits of calcium carbonate are reached. Scale then continues to form until either the carbonate species or the calcium is used up or the water chemistry changes to alter the solubility of carbonate scale. The pressure effects can be offset by temperature ones, since calcium carbonate is less soluble at higher temperature.

It takes time for the system to achieve equilibria and hence the pH and concentration of these species will change with time. Consequently, stabilising techniques may be applied to samples taken to prevent such changes from taking place until the water is analysed. These involve adding known concentrations of stabilising solutions, such as HCl, to keep the equilibria away from CO_3^{2-} and prevent calcium carbonate precipitation.

Figure 4.3 clearly illustrates the balance in relative concentration of each of the carbon species with pH. This can be used as an additional check on reported water chemistry. A low pH and carbonate concentrations higher than bicarbonate concentrations cannot be correct.

Because the organic acids such as formic, acetic, propionic, and butyric acids also contribute H^+ ions to the water chemistry, they influence the pH/CO_2/HCO_3^-/CO_3^{2-} system. Determining the concentration of the different organic acids is difficult by alkaline titration since the end points when they become neutralised are at similar pH values to each other and to bicarbonate ions. Hence ion exclusion chromatography would be the preferred technique for determining the organic acids. The bicarbonate ion concentration determined by titration (that includes the influence of the organic acids) should then be adjusted to account for the organic acids.

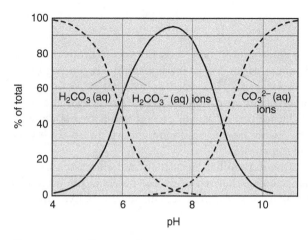

Figure 4.3 pH effect on carbonate species.

4.4.3 Monitoring Corrosion Management Strategies

In addition to determining the risk of corrosion, water samples are also frequently taken to assess the effectiveness of any corrosion management strategy, as outlined in Chapters 10 and 18. The sampling techniques are similar to those described above, although it is usually possible to sample from low pressure locations into atmospheric sample bottles. Analysis is either for the residual concentrations of corrosion inhibitors, for evidence of scale formation (loss of scaling ions) or for bacterial activity that may indicate the presence of solid deposits upstream.

4.5 Influential Parameters in System Corrosivity

The main parameters affecting system corrosivity in production steams are summarised in Table 4.3. These are dealt with in depth in Chapter 5.

Table 4.3 Influential parameters affecting system corrosivity.

Parameters and influences	Chemical parameters	Physical parameters
High influence on CO_2 metal loss corrosion of CLASs	• CO_2 content • H_2S content • In-situ pH • Chloride (for CRAs) • Organic acids and bicarbonate • Ca^{2+}/HCO_{3-} ratio • Dissolved Fe • Bacteria	Temperature Partial pressure of acidic gases Flow regime Pipe size Gas oil ratio Water content
Low influence on CO_2 metal loss corrosion of CLASs	Non-scaling ions (Na, K, etc.) Scaling ions (Ba, Sr, Mg, etc.)	System pressure

4.6 Summary

The water chemistry is critical to determining the corrosivity within production and injection systems. The dissolved anions, cations, pH, bicarbonate, and organic acids are primarily analysed to determine the scale risk. However, this information is also vital to determine the nature and magnitude of any corrosion risk. The corrosivity of the water is strongly influenced by the CO_2/bicarbonate/carbonate levels which are pH-dependent and are subject to changes in system pressure and temperature, and the presence and relative proportions of oil and gas phases.

Analysing water samples allows the determination of the water chemistry. Correct procedures are required to obtain samples that represent the waters at the particular sampling location, and to ensure that these samples remain representative until the time they are analysed.

Notes

1 http://petrowiki.org/PetroWiki.
2 Acetic acid is used here to illustrate the principle, but similar equations apply for other organic acids found in water.

References

1 Jones, L.W. (1988). *Corrosion and Water Technology for Petroleum Producers*. Tulsa, OK: OGCI.
2 Patton, C.C. (1995). *Applied Water Technology*. JMC.
3 PJ Webb, Enhanced scale management through the application of inorganic geochemistry and statistics. Paper presented at SPE International Symposium on Oilfield Scale, Aberdeen. SPE 87458-MS, 2004.
4 RT Elworthy, A field method and apparatus for the determination by means of electrical conductivity measurements the character of waters leaking into oil and gas wells. Summary report No. 605. Washington, DC: US Department of the Interior, Bureau of Mines, 1922.
5 McCain, W.D. Jr. (1990). *The Properties of Petroleum Fluids*, 2e. Tulsa, OK: PennWell Books.
6 E Butler, and P Shone, Produced water best practices: sampling guidelines, analytical methods and requirements. Paper presented at the UKOOA Condensate in Water Workshop, Aberdeen, 6–7 September, 2000.
7 Höpfner, H. (2002). *Corrosion Resistant Alloys for Oil and Gas Production: Guidance on General Requirements and Test Methods for H_2S Service, EFC 17*, 2e. London: Maney Publishing.
8 Jansen, A.J.H., Cohen Stuart, M.A., and de Keizer, A. (2005). *Biologically Produced Sulfur Particles and Polysulfide Ions*. Wageningen: Laboratory of Bionanotechnology, Wageningen University.
9 ASTM, Standard practices for water sampling. ASTM D3370–82, 1982.

10 M Yand and A Laurie, Understanding the DTI guideline notes on sampling and analysis of produced water and other hydrocarbon discharges. 4th Produced Water Workshop, 24–25 May 2006.

11 Norsk Olje and Gass, Norwegian oil and gas recommended guidelines for sampling and analysis of produced water. 085, 2013.

12 Lauer, W.C., McCandless, T.J., and Flancher, D. (2012). *AWWA Water Operator Field Guide*, 2e. American Water Works Association.

13 Wolthers, M., Van der Gaast, S.J., and Rickard, D. (2003). Structure of disordered mackinawite. *American Mineralogist* 88: 2007–2015.

14 J Boivin, Water analysis and corrosion investigations. Paper presented at NACE Northern Area Western Conference, Victoria, BC, Canada, 16–19 February, 2004.

15 JL Crolet and MR Bonis, NACE Annual Corrosion Conference, Paper No. 05272, 2005.

16 JL Crolet and J Leyer, NACE Annual Corrosion Conference, Paper No. 04140, 2004.

17 Kermani, M.B. and Morshed, A. (2003). Carbon dioxide corrosion in oil and gas production: a compendium. *Corrosion* 59: 659–683.

18 JL Crolet, N Thevenot, and S Nesic, NACE Annual Corrosion Conference, Paper No. 4, 1996.

19 JL Crolet, N Thevenot, and A Dugstad, NACE Annual Corrosion Conference, Paper No. 24, 1999.

20 JA Dougherty, NACE Annual Corrosion Conference, Paper No. 04376, 2004.

Bibliography

Burger, ED and Odom, JM, Mechanism of anthraquinone inhibition of sulfate-reducing bacteria. SPE International Symposium on Oilfield Chemistry, SPE Paper no. 50764, Houston, TX, 16–19 February, 1999.

DTI, Non-oil analysis/letter to industry/technical requirements/produced water/draft form. December 2004.

Hubert, C. and Voordouw, N.G. (2007). Oil field souring control by nitrate-reducing Sulfurospirillum spp. that outcompete sulfate-reducing bacteria for organic electron donors. *Applied and Environmental Microbiology* 73: 2644–2652.

Ostroff, A.G. (1979). *Introduction to Oilfield Water Technology*, 2e. Houston TX: National Association of Corrosion Engineers.

Penkala, JE, Shioya, N, Bastos, EC, et al. A cost effective treatment to mitigate biogenic H_2S on a FPSO. NACE Annual Corrosion Conference, Paper No. 04751, 2004.

Smith, MGD, Comparison of OSPAR analytical methods for the determination of dispersed oil in produced water. TNO Report R2003/025, February 2003.

Standards

ASTM, Standard practices for water sampling. ASTM D3370-82, 1982.

Energy Institute, Determination of the oil contents of effluent water: extraction and InftB-red spectrometric method. IP 426/98. 1998.

ISO, Water quality – sampling, Part 3: Guidance on the preservation and handling of samples. EN ISO 5667-3, 2001.

ISO, Water quality – determination of hydrocarbon oil index – Part 2: Method using solvent extraction and gas chromatography. ISO 9377-2: 2000, 2000.

NEL, Oil in water analysis method (OIWAM): a joint industry project (JIP): implementation of the ISO 9377-2 (MOD) method and the development of acceptance criteria for alternative methods. NEL Final Report No. 2005/96. 2005.

The Netherlands, Results of a 'quick scan' study: comparison of OSPAR Agreement 1997-16 with GC method based on ISO 9377-2. Paper presented at OSPAR OIC meeting, Cadiz, 11–15 February, 2002.

OSPAR, Agreement 2005-15 on sampling and analysis procedure for the 40 mg/l target standards, 2015.

5

Internal Metal Loss Corrosion Threats

The potential internal corrosion threats from handling and processing produced fluids and gases are generally more complex in nature and more aggressive to address than external corrosion. In contrast, the continuing development of coatings technology used singly and, where subsea and buried infrastructure is concerned, in tandem with the robust and well-established technology and practice of cathodic protection (CP), means external corrosion can arguably be engineered and managed more confidently than for internal corrosion. The implications of internal corrosion threats should be considered both at the design stage and to effectively and consistently manage during operating life, and should be considered far more likely to determine materials selection. However, it is the quality of application or poor provision of appropriate through-life corrosion management programme and implementation of effective corrosion mitigating measures and continual supporting maintenance that are more likely the primary causes of premature failure by internal corrosion threat.

Given the range of desirable engineering properties offered by carbon and low alloy steels (CLASs), their relative ease of fabrication, and abundance of supply, together with project economic viability considerations, mean the preference and pressure to use CLASs is invariably strong. However, CLAS has a tendency to be subject to various corrosion threats, in particular, internal metal loss corrosion.

Metal loss corrosion is the wastage of metal by electrochemical reaction with its environment that in the extreme, if left unmitigated, can lead to loss of mechanical strength and structural failure or breakdown. Its severity is governed by environmental and hydrodynamic conditions, the presence of corrosive gases, operating regimes and prevailing temperatures, in addition to metallurgical parameters. The primary source can be divided into two categories driven by dissolved gases, principally CO_2 and H_2S in production systems and O_2 in water injection systems. Confidence of being able to quantify credible internal corrosion threats and their cost-effective mitigation is therefore a critical step; and therein internal metal loss corrosion is a primary consideration.

Internal metal loss corrosion threat to CLASs is by far the most prevalent form of attack encountered in upstream hydrocarbon operation. This type of damage has long been recognised and considered an operational challenge [1–39]. Its understanding, prediction, and control are key challenges to sound facilities design, operation, and subsequent integrity assurance. CO_2/H_2S gases are usually present in produced fluids and O_2 in injection fluids causing the respective corrosion threats. While these types of

Corrosion and Materials in Hydrocarbon Production: A Compendium of Operational and Engineering Aspects, First Edition. Bijan Kermani and Don Harrop.
© 2019 John Wiley & Sons Ltd. This Work is a co-publication between John Wiley & Sons Ltd and ASME Press.

damage do not by themselves cause the catastrophic failure mode of cracking associated with H_2S, their presence in contact with an aqueous phase can, nevertheless, result in very high corrosion rates where the mode of attack is often highly localised (mesa corrosion) and hence challenging to design against.

This chapter deals with CO_2/H_2S in relation to CLASs-driven internal metal loss corrosion in upstream operations, briefly describing current understanding, means of prediction and mitigation. It is primarily a synopsis of an original article published by NACE, and reproduced with permission from NACE International, Houston, TX (all rights reserved) [1].

It should be noted that metal loss corrosion in the presence of CO_2 only is conventionally referred to as 'sweet corrosion', whereas the corrosion threat in the presence of H_2S is referred to as 'sour corrosion'.

Metal loss O_2 corrosion primarily associated with water injection systems is dealt with separately in Chapter 7.

It is worthy of note that CO_2 is some 36 times, and H_2S some 70 times, more soluble in water than O_2, but purely on a dissolved concentration basis O_2 is far more damaging in terms of corrosion rate than both CO_2 and H_2S [39] for the same concentration.

5.1 CO_2 Metal Loss Corrosion

Dry CO_2 gas by itself, like dry H_2S, is not corrosive at the temperatures encountered within the oil and gas production system. It needs to be dissolved in an aqueous phase to promote an electrochemical reaction between steel and the contacting aqueous phase. CO_2 is soluble in water and brines. However, it should be noted that it has a similar solubility in both the gaseous and liquid hydrocarbon phases. Thus, for a mixed-phase system the presence of hydrocarbon phase may provide a ready reservoir of CO_2 to partition into the aqueous phase.

5.1.1 The Mechanism

Corrosion of CLAS in CO_2-containing environments is a complex phenomenon. A number of mechanisms have been proposed for the process [1–39]. However, these either apply to very specific conditions or have not received widespread recognition and acceptance. In general, CO_2 dissolves in water to give carbonic acid, a weak acid compared to mineral acids as it does not fully dissociate (Eq. 5.1):

$$CO_2 + H_2O \Leftrightarrow CO_2 - H_2O \cong H_2CO_3 \Leftrightarrow H^+ + HCO_3^- \Leftrightarrow 2H^+ + CO_3^- \tag{5.1}$$

As a consequence of the equilibrium described in Eq. (5.1), much debate continues in the literature as to the rate-determining step (RDS) in the reaction of the dissolved CO_2 with a steel surface. These have been widely publicised over the years and covered elsewhere [1–38] and are considered outside the scope of this chapter. However, the overall corrosion reaction can simply be written as Eq. (5.2):

$$Fe + H_2CO_3 \Rightarrow FeCO_3 + H_2 \tag{5.2}$$

5.1.2 Types of Damage

Metal loss CO_2 corrosion occurs primarily in the form of general corrosion and three variants of localised corrosion: (i) pitting; (ii) mesa attack; and (iii) flow-induced localised corrosion. In studying the threat of metal loss CO_2 corrosion, a clear distinction should be made between (i) pure metal loss corrosion and (ii) combined erosion/metal loss corrosion where the interaction is commonly synergistic – i.e. the actual metal loss experienced is greater than purely the sum of the erosion rate and corrosion rate acting alone with the resulting damage morphology often localised. The latter characterises itself in the form of ripple marks, horseshoe, comet tails and dinosaur footprints, whereas the former is as described briefly in this chapter. Three variants of metal loss CO_2 corrosion are outlined herewith.

5.1.2.1 Pitting

Pitting more often occurs at low velocities and around the dew-point temperatures in gas-producing wells. In the field, the presence of pits may be the result of an upset in or subtle changes to operating conditions, adjacent to non-metallic inclusions, in the vicinity of welds and associated with incipient mesa attack. Pitting susceptibility increases with temperature and CO_2 partial pressure.

The discussion on pitting of CLASs in sweet environments remains a continuing focus of research and debate. Various authors attribute pitting initiation and its propagation to different factors and there is generally no applicable rule for its prediction.

5.1.2.2 Mesa-Type Attack

Mesa attack is a type of localised corrosion and occurs in low-to-medium flow conditions where the protective iron carbonate film (scale) forms but it is unstable to withstand the operating regime or fails, due to harsh hydrodynamic conditions generated, often locally and maybe transitory, as a result of the nature of the system hydrodynamics or fluid flow regime. It manifests itself in large flat bottom steps with sharp edges. Corrosion damage in these locations is significantly in excess of the surrounding areas.

Again, the exact conditions under which mesa attack forms is a source of continuing debate and research to prevent with absolute certainty its occurrence in the field in the absence of continuous application of any corrosion mitigation programme, e.g. treatment with a corrosion inhibitor which can be very effective, subject to selection of the right inhibitor always present at its optimum concentration.

5.1.2.3 Flow-Induced Localised Corrosion

This form of corrosion typically starts from pits or sites of mesa attack exposed to above critical flow intensities. It then propagates by local turbulence created by the pits or steps at the mesa attack or by protruding geometry and in the region of bends – e.g. high velocity lines are potentially subject to pitting at the outside of bends or downstream of features, such as weld beads; whereas low velocity lines potentially are subject to pitting or channelling at bottom-of-line. The local turbulence combined with the stresses produced during further scale growth may destroy the existing corrosion scales. Once the corrosion scale is damaged or destroyed, the flow conditions may then prevent reformation of protective scale on the exposed metal.

5.2   Key Influential Factors

Metal loss CO_2 corrosion is influenced by a number of parameters including: (i) environmental; (ii) physical; and (iii) metallurgical variables, as illustrated in Figure 5.1. The majority of these have been extensively covered by a number of authors and captured over the past decades. The key point of consideration in assessing the influence of these parameters is their respective influence on the formation of a stable protective $FeCO_3$ film (scale). This film and FeS film formed in the presence of H_2S both may offer protection and ensure a subsequent lowering of corrosion damage. A thin highly protective FeS film can form very rapidly, even in the presence of very low levels of H_2S. However, as FeS is conductive, any localised breakdown of this can drive high pitting rates due to the formation of a galvanic cell between the intact sulphide film (the large cathode) and the point of localised breakdown (the small anode) [12]. The subject is extremely complex and the uncertainties remaining in uniquely defining the interaction of all the prevailing parameters affecting internal metal loss CO_2 corrosion continue to raise challenges in the ability to always confidently predict type and/or rate of damage on CLAS components, unequivocally and/or quantitatively.

In brief, notable parameters affecting metal loss CO_2 corrosion include:

- fluid make-up as affected by water chemistry, pH, organic acids, water wetting, hydrocarbon characteristics, and phase ratios;
- CO_2 and H_2S contents;

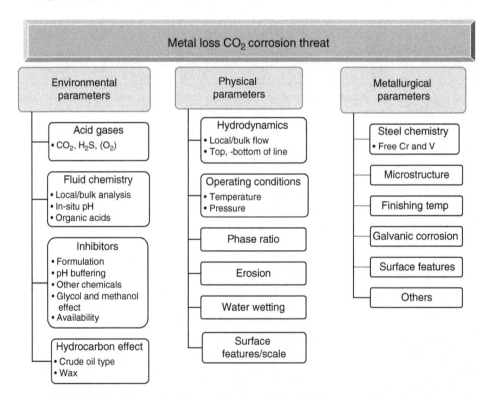

Figure 5.1 Metal loss CO_2 corrosion threat, the influential parameters – also applicable to metal loss CO_2/H_2S corrosion threat.

- temperature;
- steel surface, including corrosion film morphology, presence of wax and ashphaltene;
- fluid dynamics/flow regime;
- steel chemistry.

All parameters are interdependent and can interact in many ways to influence metal loss CO_2 corrosion. The three principal parameters affecting metal loss CO_2 corrosion and relative influence of their respective constituents on film formation and growth are summarised in Table 5.1. While the majority of these parameters are dealt with elsewhere with some notable reference in this chapter, the topic of inhibition is covered in Chapter 8.

5.2.1 Notable Parameters

While the majority of parameters in Table 5.1 and Figure 5.1 are important in influencing the corrosion process, they have been dealt with in depth in many publications over the past few decades [1–39]. Here only two parameters are described briefly as they constitute notable examples of parameters affecting metal loss CO_2 corrosion process of CLASs.

5.2.1.1 The In-Situ pH
Solution pH plays an important role in the corrosion of CLASs [1] by influencing both the electrochemical reactions that lead to iron dissolution and the precipitation of protective scales that govern the various transport phenomena associated with the former. Under certain conditions, solution constituents of the aqueous phase will buffer the pH which can lead to precipitation of a corrosion scale and possible lowering of corrosion rates. It should be noted that, in certain circumstances, the resulting corrosion product can be corrosive and increase the severity of attack [1–3, 20, 21].

5.2.1.2 The Effect of Organic Acid
Organic acids present in production fluids have long been considered to significantly influence and complement the severity of CO_2 corrosion. The influence has been shown to occur systematically in all field conditions where CO_2 corrosion was observed [1, 21–23]. The presence of acetic acid (HAc) in production streams, and particularly gas-producing fields, reduces the protectiveness of the films and increases the sensitivity to mesa attack. This is attributed to a lower Fe^{2+} supersaturation in the corrosion film and at the steel surface [5, 20, 24–26].

Generally the presence of acetic acid causes a significant increase in the corrosion rates in CO_2 environments [23, 25]. Acetic acid (along with other organic acids) can jeopardise the protective corrosion product scales formed in top-of-line corrosion [32].

At low CO_2 partial pressure, CO_2 corrosion disappears, but in certain fields, it can be replaced by a genuine 'acetic acid corrosion'. It has been shown that this was not due to any influence of the acetic acid, either on the cathodic reaction of H^+ or on the anodic dissolution of iron, but rather to its effect on the protectiveness of the corrosion product layer formed. In the presence of traces of free acetic acid, the majority of corrosion product formed on bare metal was no longer iron carbonate, but iron acetate, which had a much greater solubility [1, 20].

Table 5.1 Significance of operational parameters in affecting metal loss CO$_2$ corrosion.

Parameters	Influence on metal loss CO$_2$/H$_2$S corrosion severity of CLAS								
Environmental	In-situ pH	Chloride content	Supersaturation	Salinity	pCO$_2$	pH$_2$S	Organic acid	CO$_2$/H$_2$S ratio	Ca^{++}/HCO$_3^-$ ratio
	S	L	S	L	S	S	S	S	S
Physical	Temp	Total pressure	Surface roughness	Fluid dynamics	Surface characteristics (prior scale)	Stress	Water wetting	Crude oil	
	S	L	M	M	M	L	S	M	
Metallurgy and composition of CLAS	Cr	C	Mn	V	Si	S/P	Mechanical properties	Heat treatment	Microstructure
	S	M	L	S	L	M	L	M	S

S: Strong influence.
M: Medium influence or variable influence.
L: low or no influence.

In general terms, organic acids affect Fe_{sat} solubility and supersaturation limits and hence interfere with the formation of a protective $FeCO_3$ layer. This has a strong influence on CO_2 corrosion and hence affects corrosivity assessment and prediction [1]. The effect of organic acid on FeS films is less clear as it is less soluble, i.e. it requires a much lower concentration of dissolved Fe^{2+}. As indicated here and also in Chapter 4, organic acids are generally more prevalent in gas systems and are a key factor with respect to top-of-line (TOL) corrosion; but the presence of acetate in all sources of produced water should be analysed for such is its potential to exacerbate CO_2 corrosion rate above that predicted in its absence.

It has been reported that at a given pH [23], any replacement of a concentration or a flux of bicarbonate by an equivalent quantity of acetate would considerably increase the local solubility of iron. This decreases the protectiveness of the corrosion product formed by increasing iron concentration gradients in solution, i.e. it suppresses the formation and stability of a corrosion product surface film.

5.3 Metal Loss CO₂ Corrosion Prediction

Over the years much effort has been expended on studying factors controlling the performance of CLASs in production environments in an attempt to define their safe operating limits in terms of the environmental and physical conditions. This knowledge has been gained, in part, from: (i) comprehensive empirical-based field statistics, such as those by API; (ii) laboratory-based information translated by quantitative regression to predict a corrosion rate; and (iii) comprehensive field statistics based upon the actual field values of relevant scientific quantities. There are also increasing attempts and interest directed at mechanistic modelling, notably the extensive and continuing work undertaken at Ohio University through a long-running oil industry-sponsored programme that has developed the CO_2 and H_2S mechanistic model Multicorp [16, 18, 19]. Nevertheless, it should be noted that some of these models are not widely available.

Having determined the water chemistry, this then needs to be translated into a corrosion risk, preferably expressed quantitatively as a rate and type of attack. The easiest parameter to relate to (and in general to measure) experimentally and in the field is the partial pressure of CO_2. Thus, over the years, a number of empirical relationships have evolved with varying levels of complexity and theoretical substance. In the simplest form, API in the late 50's provided 'rule of thumb' criteria, where partial pressure of CO_2 is denoted as Pco_2 for CLASs [32] as follows:

- $Pco_2 < 0.5$ bar (7 psi)
  Corrosion unlikely; implying that corrosion is uniform and the rate is below 0.1 mmy⁻¹.
- 0.5 bar (7 psi) < Pco_2 < 2 bar (30 psi)
  Corrosion possible; implying that corrosion rate may be between 0.1–1 mmy⁻¹ and design consideration is based on the life expectancy.
- $Pco_2 > 2$ bar (30 psi)
  Corrosion likely; implying that corrosion rate may exceed 1 mmy⁻¹ and this is unacceptable.

These rules are based on field experience, principally in the US and still have an engineering use. No subsequent standard guidelines exist as to the course of action necessary for each condition, although experience will trigger the corrosion engineer to initially look at certain possible options (e.g. use of a corrosion inhibitor or more corrosion-resistant alloys). However, designing on such a basis is in itself not a very satisfactory way to operate.

Rules of thumb may be viewed as:

- an aid to first-pass materials selection;
- offering qualitative and generalised assessment.

More specific questions on service life, risk analysis, corrosion allowance, and inhibited corrosion rates require the ability to conduct a quantitative assessment, i.e. the ability to predict corrosion rates for a given set of conditions. In addition, the consequences of corrosion are a key issue when a corrosivity assessment is carried out. Nevertheless, the relationship between the corrosion rate and CO_2 partial pressure has formed the basis of a number of predictive models/equations for carbon steels with varying degrees of general applicability, depending on their empirical origins versus mechanistic detail and sophistication.

There are a number of publications outlined in the Bibliography that specifically address a comparison of different models and their strengths and weaknesses.

5.3.1 Industry Practice

In addressing metal loss corrosion, it is paramount to bear in mind that corrosion is a multi-disciplinary process [39] and therefore depends on influences brought by the material, the environment it is exposed to, and the circumstances of exposure. Such a combination of parameters is not easily conducive to producing a wholly precise deterministic predictive tool. Nevertheless, the primary mechanisms of the most common forms of metal loss attack, and particularly 'sweet corrosion', have been extensively studied over many years and increasingly detailed. It is therefore, the responsibility of Subject Matter Experts to use this knowledge to provide balanced guidance on standard use of predictive models by an organisation for both design and operational application. As yet there is not, and may never be, a single 'right' model.

Over the years, many CO_2 corrosion prediction models have been developed by the operators, institutions, and service companies in an attempt to assess potential corrosivity as affecting the performance of CLASs, all claiming credibility [4–7, 10, 11, 13–19, 26–28, 30, 33]. Due to the complexity of the corrosion process and the philosophies used to develop the models, different outcomes can be derived when deploying these models. These models rely on contrasting and influential parameters, including protection by corrosion product films, water/oil wetting, organic acids, in-house field experience, and others. In the majority of cases, the models have a built-in conservatism and may over-predict the corrosion attack. There is a trade-off between a model's relative ease of use versus availability and extent of detail and reliability and the accuracy of the input data and the conditions required. Furthermore, the accuracy of the predicted rate may be more a result of how the model has been set up to compute it – e.g. output to two decimal places – than actually reflecting the inherent accuracy of the model. Looking to draw a correlation between predicted rates and documented rates from field analogues can help here, whereas flipping between models is really the domain of a Subject Matter Expert.

While an industry-standard approach to predict metal loss CO_2 corrosion damage does not exist per se, the work of de Waard et al. [10, 11, 14] at Shell, latterly formulated into Hydrocor [15, 16], has provided a strong reference statement from or against which to work. The resulting primarily empirical equations and nomograms have been developed as an engineering tool. It presents in a simple form the relationship between 'potential corrosivity' (worst case) of aqueous media for a given level of dissolved CO_2, defined by its partial pressure, at a given temperature. The relative simplicity of the de Waard et al. approach and ease of use have undoubtedly been positive factors in its broad acceptance, although its complexity has grown over the years as better appreciation of the metal loss CO_2 corrosion process and the influence of water chemistry, surface films, and fluid dynamics have evolved together with better correlation with field experience [4, 7, 34, 35].

On the other hand, the well database contained within the spreadsheet called 'CORMED' developed by Crolet and Bonis [4, 8, 9] aims to translate the analytical sheets of the raw field data into factual inputs (in-situ pH, free HAc, etc.) and automatically compare them to their respective critical values in published tables. CORMED places direct emphasis on knowledge of the water chemistry and defines severity in terms of low, medium or high risk. CORMED and Lipocor [34] have been subsequently incorporated into an inclusive program called Corplus and latterly PreCorr. It is a deterministic tool for predicting the potential corrosivity in oil and gas production to determine the in-situ pH and predict corrosion rate. This is used for design purposes, material selection, and, to some extent, likelihood of damage.

An alternative model provided by Norsok (Norsok M-502 Corrosion Rate Calculation Model [13]), a version of which is freely available online, is widely used by the industry. It contains a computer program to calculate CO_2 metal loss corrosion rate. It is complemented by a manual on how to use the model, its limitations, and the necessary inputs and outputs. The model incorporates the majority of de Waard algorithms complemented by advances made in flow modelling and further understanding of metal loss CO_2 corrosion.

Cassandra [26–28] by BP differentiates itself by covering topics, such as the calculation of pH, the treatment of fugacity, scaling, oil wetting, acetate, and hydraulic diameter. In addition, there is the ECE [33] model that can be purchased and is used by smaller operators.

All models, including de Waard nomograms, CORMED/PreCorr approach, and Norsok, have been developed from a basic consideration of the metal loss CO_2 corrosion reactions. de Waard, ECE, and Norsok are mainly empirical in origin while CORMED/PreCorr are more theoretical. These have then attempted to account for the underlying effects either by applying correction factors (de Waard and Norsok) or through field correlation (CORMED/PreCorr).

Recently, Multicorp [6, 18, 19], developed by Ohio University through a multi-sponsored programme, is a corrosion prediction engine that claims simulation of corrosion under various conditions and in various environments. Multicorp is based on a mechanistic (theoretical) model, reflecting faithful descriptors of the important physico-chemical processes underlying corrosion.

Finally, it is worth restating that all available models/equations only apply to CLASs and, with the exception of the Ohio University Multicorp mechanistic model, need to be used with caution in the presence of H_2S due to the formation of protective sulphide films which may nevertheless be susceptible to localised breakdown under long-term exposure.

It should be reiterated that the above models are all developed to predict CO_2 metal loss corrosion of CLASs. There are a few limited publications in relation to metal loss corrosion of 13%Cr.

However, notwithstanding all the above, predictive models have come to be an essential tool for the corrosion engineer to use both during project design and throughout a field's operational life in the conduct of corrosion risk assessments and to assess the significance of corrosion monitoring and inspection data. The model to be used is often governed by company/operator preference, with care needing to be exercised in the hands of a casual user. It is important to have a working understanding of the origin of the model to be used and how it addresses the key factors that determine the predicted corrosion rate (e.g. applicable partial pressure of CO_2 and temperature range; $FeCO_3$ protective scale formation; in-situ pH; presence of H_2S and acetate; influence of flow regime). Flipping between the uses of the various models is really the domain of the Subject Matter Expert (SME).

5.4   Metal Loss Corrosion in Mixed H_2S/CO_2 Containing Streams

Ignoring the cracking aspects of corrosion problems associated with sour service, low levels of hydrogen sulphide can affect CO_2 corrosion in different ways. The presence of H_2S affects materials and in particular CLAS in a similar manner to that of CO_2 with all influential parameters outlined earlier for metal loss CO_2 corrosion affecting its process and mechanism (Figure 5.1). H_2S can either increase CO_2 corrosion by acting as a promoter of anodic dissolution through sulphide adsorption and affecting the pH or decrease sweet corrosion through the formation of a protective sulphide scale [12, 29, 36, 37].

Many papers have been published on the interaction of H_2S with CLASs. However, literature on the interaction of H_2S and CO_2 is more limited since the nature of the interaction is complex. The corrosion reaction often leads to the formation of iron sulphide (FeS) scales, which, under certain conditions, rapidly form and are highly protective. However, their breakdown (e.g. under highly turbulent flow conditions or due to erosion in the presence of solids such as sand) can lead to very severe localised corrosion in a similar manner to that for $FeCO_3$ breakdown in the case of CO_2 corrosion alone. However, the resulting localised rates in the presence of H_2S can be much higher due to the conductive nature of the remaining intact FeS filmed areas (cf. large cathode/small anode).

Whether this mechanism predominates over that of CO_2 attack, which is also present, will depend on the circumstance and the relative levels of CO_2 and H_2S present (see Table 5.1). In many cases, CO_2 corrosion dominates but the rate of attack is modified in the presence of H_2S. In such circumstances, generally corrosion rates are lower but the risk of localised attack increases. The kinetics and nature of FeS film formation, stability, and its contribution to reducing corrosion are key to affording protection.

As a general rule in CO_2-containing environments, the presence of H_2S can:

- Increase the corrosion risk by either:
  - facilitating localised corrosion, at a rate greater than the general metal loss or localised rate expected from CO_2 corrosion alone; or
  - preferentially forming a weakly protective FeS corrosion product particularly at low levels of H_2S ahead of and interfering with establishing favourable conditions for less protective iron carbonate formation [8, 9].
- Decrease the corrosion risk by promoting the formation of an FeS corrosion product film through either
  - more readily forming a stable protective FeS film for a given level of dissolved Fe^{2+} in solution than iron carbonate replacing a less protective iron carbonate film [8, 9]; or
  - forming a combined protective layer of iron sulphide and iron carbonate.

A more detailed overview of types and nature of films formed on CLASs by H_2S in CO_2-containing environments can be found elsewhere [8, 9, 29].

5.4.1 Assessment Methods

While there is no internationally recognised governing rules to assess potential system corrosivity, in the presence of both acid gases, the corrosion process is governed by the dominant acid gas. The presence of H_2S in CO_2-containing producing environments has been reviewed by Pots [16], and Bonis [37] and Joosten [29]. Pots introduced the notion of a CO_2/H_2S ratio to define three corrosion domains based on the prevailing dominant corrosion mechanism. Pots' approach is presented in Table 5.2.

Recent practical experience in a large number of fields reported by Bonis [17, 37] has characterised sour fields into three corrosion severity categories with three distinct corrosion mechanisms: (i) negligible; (ii) moderate; and (iii) very severe. The incidence of severe cases was reported to be very rare. No relationship with potential system corrosivity predicted by the usual CO_2 corrosion models was found. It is shown that flow velocity and flow regimes are the most influential factors for transition between the categories. It is also highlighted that very severe corrosion requires 'pit promoters' such as sulphur, oxygen, bacteria, and others and a 'galvanic effect' with surrounding non-corroding surfaces. It appears from reviewed experience that these promoters are

Table 5.2 Dominance of metal loss CO_2 corrosion by CO_2/H_2S ratio.

CO_2/H_2S	Corrosion dominated by:	Surface film/scale	Corrosion prediction tool
<20	H_2S metal loss corrosion	FeS as the main corrosion product	None available
Between 20 to 500	Mixed CO_2/H_2S metal loss corrosion	A mixture of FeS and $FeCO_3$ as the main corrosion products	Limited predictive models available – potentially unproven field backup
>500	CO_2 metal loss corrosion	$FeCO_3$ as the main corrosion product	CO_2 Corrosion Models

Source: [37, 38].

mostly extraneous to produced fluids, apart from sulphur depositing from sour gases. The basic mechanism of cathodic and anodic insoluble layers is reiterated as the key mechanism affecting the transition between the three categories.

While Pots [16] categorised H_2S/CO_2 metal loss corrosion into three regimes tabulated in Table 5.2, Bonis [37] has highlighted that H_2S and CO_2 partial pressure, pH or the H_2S/CO_2 ratio do not influence the corrosion likelihood in any significant way so long as a minimal amount of H_2S is present above a CO_2/H_2S ratio of 20. Other parameters, including water salinity and temperature, are reported to be less effective. It is also noted that most factors have a similar corrosion contribution in oil as in gas production systems, in wells and pipelines. This suggests that the basic corrosion mechanisms involved within these different facilities are not vastly different.

Where metal loss is the primary problem, the use of a corrosion inhibitor can be an effective control measure; indeed some inhibitors appear to work better in the presence of H_2S or more generally sulphide – the addition of sodium or ammonium sulphide to an inhibitor formulation is not uncommon, certainly in the past.

5.5 Summary

This chapter has led to a number of key conclusions emphasising the importance of the subject in providing integrity management for oil and gas production facilities. These include:

- Metal loss CO_2 corrosion of CLASs is a complex phenomenon, the understanding and prediction of which have benefitted substantially over the past 60 years from significant empirical and mechanistic study in the laboratory and field. This has resulted in the development of a number of predictive models, each of which has its strengths and weaknesses and varying degrees of 'sophistication and complexity'. However, there continues to be limited appetite for establishing an industry-endorsed and industry-adopted single model as a standard to at least start with.
- Metal loss CO_2/H_2S corrosion is influenced by a large number of parameters, including environmental, physical, and metallurgical variables. All parameters are interdependent and can interact in many ways to influence metal loss CO_2 corrosion.
- Ignoring the corrosion problems associated with corrosion cracking in sour service, hydrogen sulphide can affect metal loss CO_2 corrosion in different ways either complementing metal loss CO_2 corrosion by acting as a weak acid, through the formation of a protective sulphide scale or, as the ratio increasingly favours H_2S, introducing the potential threat of pitting.
- The presence of acetic acid or more generally organic acids reduces the protectiveness of $FeCO_3$ films and increases the sensitivity to mesa attack. This is attributed to a lower Fe^{2+} supersaturation in the corrosion film and at the steel surface in the presence of organic acids.
- Corrosion scales (primarily $FeCO_3$), when formed under certain conditions, can afford superior protection. While their formation and growth have been the subject of many studies, favourable conditions for the development of a truly protective film to provide subsequent reliable/resilient long lasting effective protection require further scrutiny.
- Steel chemistry plays a significant role in providing protection against metal loss CO_2 corrosion and can lead to substantial economic gains.

References

1 Kermani, M.B. and Morshed, A. (2003). Carbon dioxide corrosion in oil and gas production. *Corrosion* 59: 659–683.

2 Kermani, M.B. and Smith, L.M. (eds.) (1994). *Predicting CO_2 Corrosion in the Oil and Gas Industry* Publication No. 13. European Federation of Corrosion.

3 Kermani, M.B. and Smith, L.M. (eds.) (1997). *CO_2 Corrosion Control in Oil and Gas Production – Design Considerations*. Publication No. 23. European Federation of Corrosion.

4 Crolet, J.L. and Bonis, M.R. (1991). Prediction of the risks of metal loss CO_2 corrosion in oil and gas wells. *SPE Production Engineering* 6 (4): 449–453.

5 JL Crolet and MR Bonis, The role of acetate ions in metal loss CO_2 corrosion. NACE Annual Conference, Paper No. 160, 1983.

6 S Nešić, H Li, J. Huang, and D Sormaz, An open source mechanistic model for CO_2 / H_2S corrosion of carbon steel. NACE CORROSION/09 Conference, Paper No. 09572, 2009.

7 R Nyborg, Overview of metal loss CO_2 corrosion models for wells and pipelines. NACE Annual Corrosion Conference, Paper No. 02233, 2002.

8 JL Crolet and M Bonis, Algorithm of the protectiveness of corrosion layers 2 – protectiveness mechanisms and CO_2 corrosion prediction. NACE Annual Corrosion Conference, Paper No. 10363, 2010.

9 JL Crolet and M Bonis, Algorithm of the protectiveness of corrosion layers 2 – protectiveness mechanisms and H_2S corrosion prediction. NACE Annual Corrosion Conference, Paper No. 10365, 2010.

10 C de Waard, U Lotz, and A Dugstad, Influence of liquid flow velocity on metal loss CO_2 corrosion: a semi-empirical model. NACE Annual Corrosion Conference, Paper No. 128, 1995.

11 de Waard, C., Lotz, U., and Milliams, D.E. (1991). Predictive model for metal loss CO_2 corrosion engineering in wet natural gas pipelines. *Corrosion* 47 (12): 976–985.

12 Gunaltun, Y. (ed.) (2017). *CO_2 and H_2S Metal Loss Corrosion – 10 Year Review*. Houston, TX: NACE International.

13 NORSOK, Norsok M-502 corrosion rate calculation. http://www.standard.no/ petroleum (accessed 3 October 2018).

14 C de Waard and U Lotz, Prediction of metal loss CO_2 corrosion of carbon steel. NACE Annual Corrosion Conference, Paper No. 69, 1993.

15 B F M Pots, Mechanistic models for the prediction of metal loss CO_2 corrosion rates under multi-phase flow conditions. NACE Annual Corrosion Conference, Paper No. 137, 1995.

16 F M Pots, RC John, IJ Rippon, et al., Improvements on de Waard corrosion prediction and applications to corrosion management. NACE Annual Corrosion Conference, Paper No. 02235, 2002.

17 M R Bonis and JL Crolet, Basics of the prediction of the risks of metal loss CO_2 corrosion in oil and gas. NACE Annual Corrosion Conference, Paper No, 466, 1989.

18 MULTICORP, Institute for Corrosion and Multiphase Technology, Ohio University. https://www.ohio.edu/engineering/corrosion/ (accessed 3 October 2018).

19 S Nešić, Si Wang, J Lee, et al., A new updated model of CO_2/H_2S corrosion in multiphase flow. NACE Annual Corrosion Conference, Paper No. 08535. 2008.

20 JL Crolet, N Thevenot, and A Dugstad, Role of free acetic acid on the CO_2 corrosion of steels. NACE Annual Corrosion Conference, Paper No. 24, 1999.

21 Crolet, J.L. (2002). Corrosion in oil and gas production. In: *Corrosion et Anticorrosion* (ed. G. Béranger and H. Mazille). Paris: Hermès Science.

22 Crolet, J.L. (1994). Which CO_2 corrosion, hence which prediction? In: *Predicting CO_2 Corrosion in the Oil and Gas Industry* (ed. M.B. Kermani and L.M. Smith) Publication No. 13. European Federation of Corrosion.

23 Y Garsany, D Pletcher, and B Hedges, The role of acetate in CO_2 corrosion of carbon steel: has the chemistry been forgotten?, NACE Annual Corrosion Conference, Paper No. 02273, 2002.

24 J D Garber and KA Sangita, Factors affecting iron carbonate scale in gas condensate wells containing CO_2. NACE Annual Corrosion Conference, Paper No. 19, 1998.

25 JL Crolet, S Olsen, and W Wilhelmsen, Influence of a layer of indissolved cementite on the rate of the CO_2 corrosion of carbon steel. NACE Annual Corrosion Conference, Paper No. 4, 1994.

26 B Hedges, R Chapman, D Harrop, et al., A prophetic CO_2 corrosion tool – but when is it to be believed? NACE Annual Corrosion Conference, Paper No. 05552, 2005.

27 Z Zhand, S Hernandez, R Woollam, et al., CO_2 corrosion prediction along the length of a flow line. NACE Annual Corrosion Conference, Paper No. 11068, 2011.

28 R Woollam, J Vera, C Mendez, and P Echegoyen, Steady state CO_2 corrosion – a novel testing approach. NACE Annual Corrosion Conference, Paper No. 11075, 2011.

29 M Joosten and S Smith, Corrosion of carbon steel by H_2S in CO_2 containing environments – 10 year review. NACE Annual Corrosion Conference, Paper No. 5484, 2015.

30 C de Waard, L Smith, and B Craig, Influence of crude oil on well tubing corrosion rates. Paper presented at Eurocorr 2001, Riva del Garda, Italy, 2001.

31 B F M Pots and E L J A Hendriksen, CO_2 corrosion under scaling conditions: the special case of top-of-line corrosion in wet gas pipelines. NACE Annual Corrosion Conference, Paper No. 31, 2000.

32 American Petroleum Institute (1958). *Corrosion of Oil and Gas Well Equipment*. Dallas, TX: American Petroleum Institute.

33 L Smith and C deWaard, Corrosion prediction and materials selection. NACE Annual Corrosion Conference, Paper No. 05648, 2005

34 Y M Gunalton, Combining research and field data for corrosion rate prediction. NACE Annual Corrosion Conference, Paper No. 14, 1994.

35 J Kolt, E Buck, D Erickson, and M Achour, Corrosion prediction and design considerations for internal corrosion in continuously inhibited wet gas pipelines. Paper presented at UK Corrosion Conference 90, 1990.

36 B Kermani, JW Martin, and K Esaklul, Materials design strategy: effects of H_2S/CO_2 corrosion on materials selection. NACE Annual Corrosion Conference, Paper No. 06121, 2006.

37 M Bonis, Weight loss corrosion with H_2S: from facts to leading parameters and mechanisms. NACE Annual Corrosion Conference, Paper No. 09564, 2009.

38 Jones, L.W. (1988). *Corrosion and Water Technology for Petroleum Producers*. Tulsa, OK: OGCI Publications.

39 Crolet, J.L. (2016). Analysis of the various processes downstream cathodic hydrogen charging, I: Diffusion, laboratory permeation and measurement of hydrogen content and diffusion coefficient. *Matériaux & Techniques* 104 (205).

Bibliography

American Petroleum Institute (1958). *Corrosion of Oil and Gas Well Equipment*. Dallas, TX: American Petroleum Institute.

Gunaltun, Y. (ed.) (2017). *CO_2 and H_2S Metal Loss Corrosion – 10 Year Review*. NACE International.

Hausler, R.H. and Giddard, H.P. (eds.) (1985). *Advances in Metal Loss CO_2 Corrosion*, vol. I. NACE International.

Hausler, R.H. and Giddard, H.P. (eds.) (1986). *Advances in Metal Loss CO_2 Corrosion*, vol. II. NACE International.

Kermani, B, Depiction of metallurgical parameters as governing CO_2 corrosion. NACE Annual Corrosion Conference, Paper No. 3813, 2014.

Lotz, U, van Bodengom, L, and Ouwehand, C, Effect of oil or gas condensate on carbonic acid corrosion. NACE Annual Corrosion Conference, Paper No. 41, 1990.

Nyborg, R, CO_2 corrosion models for oil and gas production systems. NACE Annual Corrosion Conference, Paper No. 10371, 2010.

Sun, W., Nešić, S., and Woollam, R.C. (2009). The effect of temperature and ionic strength on iron carbonate ($FeCO_3$) solubility limit. *Corrosion Science* 51: 1273–1276.

Zheng, Y., Brown, B., and Nešić, S. (2014). Electrochemical study and modeling of H_2S corrosion of mild steel. *Corrosion* 70 (4): 351–365.

6

Environmental Cracking (EC)

Environmental cracking (EC) threats can occur when a combination of metallurgical, mechanical, and environmental conditions combine. In the majority of cases, these are specific to an alloy/environment system, although not necessarily. The threats can be catastrophic and hence require extreme care in early design, joining, and installation of components and facilities.

EC threat in hydrocarbon production manifests itself in several forms primarily associated with a combination of fluid chemistry including in-situ pH, metallurgical, and mechanical status, and operating temperatures. Principally, these forms of cracking can be divided into three groups driven by H_2S, chloride or a combination of the two. The presence of H_2S can lead to threat types driven by hydrogen-assisted cracking mechanisms, including sulphide stress cracking (SSC) and hydrogen-induced cracking (HIC)/stepwise cracking (SWC), and derivatives thereof. The presence of chloride raises the possibility of chloride stress corrosion cracking (Cl⁻SCC) but also can increase the susceptibility and severity of SSC.

This chapter outlines the most prevalent types of EC threats in hydrocarbon production and the measures used to mitigate their occurrence. As the subject area has been exhaustively addressed over the years, little or no emphasis on their respective mechanisms is made here. Major references are made to international standards as the precursor to materials design. It is important to note that other types of corrosion threat associated with the presence of H_2S and chloride, such as pitting or crevice corrosion should be considered at the design stage but these are not specifically addressed in this chapter.

Due to the significance of ISO 15156 standard in addressing H_2S-related corrosion cracking threats in hydrocarbon production, a summary of the standard is included. The whole issue of materials selection for oil and gas production and transportation is a major topic in itself which is covered in Chapter 14.

6.1 Environmental Cracking Threat in Steels

EC affects metallic materials in a number of ways depending on their physical metallurgy and structure. As discussed in Chapters 2 and 3, the metallurgical structures of steels are affected by alloying constituents and in particular Cr and Ni as shown schematically

Corrosion and Materials in Hydrocarbon Production: A Compendium of Operational and Engineering Aspects, First Edition. Bijan Kermani and Don Harrop.
© 2019 John Wiley & Sons Ltd. This Work is a co-publication between John Wiley & Sons Ltd and ASME Press.

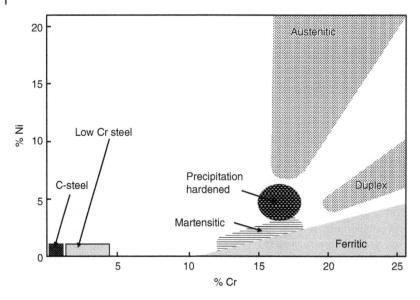

Figure 6.1 Metallurgical microstructures of corrosion resistant alloys as affected by Cr and Ni contents.

in Figure 6.1. These in turn govern the mechanism by which different metallurgical microstructures are affected by aspects of the EC threat, which are summarised in Table 6.1. While Table 6.1 provides a general overview of the governing EC threat types, it should be used with caution due to the interactive nature of metallurgy, stress, environment, and particularly the so-called elastic region which may elude some microstructures and in particular corrosion-resistant alloys (CRAs).

Apart from SSC, Cl-SCC, and SWC/HIC, other potential forms of cracking that may occur include stress-orientated hydrogen-induced cracking (SOHIC), soft zone cracking (SZC), and galvanically-induced hydrogen stress cracking (GHSC). While some of these damage threats are discussed in the present chapter, their effective roles are beyond the scope of the publication and are dealt with in the literature. The types of EC threat in relation to hydrogen entry and transport are briefly defined in Section 6.2, including the cause and manifestation, susceptible locations, and potential mitigating measures.

Assessment of respective types of threat is an essential element of qualification and quality control as discussed in Chapters 1 and 5.

6.2 EC Associated with Hydrogen Sulphide

While CO_2 metal loss corrosion is the most prevalent form of attack associated with oil and gas production and transportation, the presence of H_2S presents a more threatening type of corrosion damage. Apart from being associated with causing general or localised types of corrosion damage under favourable conditions (see Chapter 5), H_2S can facilitate catastrophic failures or cracking for which precautionary measures are essential, and care in the selection of appropriate materials is paramount. The service conditions in which cracking may threaten integrity and hence require metallurgical design

Table 6.1 Typical environmental cracking (EC) threat types in upstream hydrocarbon operations.

Metallic microstructure/ parameter	Potential temperature for maximum propensity to EC threat					
	Ambient temperatures (typically <60 °C)	Elevated temperatures typically >100 °C	Medium temperatures (typically between 60 and 100 °C)	Potentially throughout operating temperature conditions		
				GHSC	SWC	SOHIC
Potential cracking mechanism	SSC	CI-SCC	Combination of SSC and CI-SCC^a	GHSC	SWC	SOHIC
Primary driver	H$_2$S	Chloride	Combination of chloride and H$_2$S	Combination of chloride, H$_2$S and galvanic coupling	H$_2$S	H$_2$S (potentially over protection by CP)
Medium to high strength CLASs (typically seamless and higher than API 5CT grade L80)	P	N/A	N/A	N/A	N/A	N/A
Low to medium strength CLASs (typically flat products and lower grades than API 5 L Grade X65)	N/A	N/A	N/A	N/A	P	P2
Martensitic	P	S	N/A	S	N/A	
Lowly alloyed martensitic	P	P	N/A	S		
Ferritic	P	N/A	N/A	P		
Austenitic	S	P	S	S		
Duplex	S	P	P2	S		

^{a} It is to be noted that for the occurrence of classical SCC, a combination of microcreep and a strain-sensitive passive layers is required. For solution annealed austenitic structures, microcreep exhibits both a temperature threshold of around 60 °C and a stress threshold. However, cold worked duplexes may not show such a threshold and all their resistance is interrelated to many parameters, including temperature, chloride content, level of cold work, stress level, and yield strength.
P: Indicates primary EC mechanism.
P2: Indicates potential EC.
S: Indicates secondary EC mechanism.
N/A: Not applicable.

or operational precautions are generally termed 'sour service'. This is in contrast to 'sweet service' associated with CO_2-containing media where no metallurgical design or operational precautions are normally required in order to avoid EC. As discussed in Chapter 5, metal loss corrosion is the principal damage mechanism driven exclusively by CO_2 or acting in association with the presence of H_2S.

6.2.1 Corrosion Implications and Mechanism

The presence of H_2S has several key implications for operational activities including:

- It is an extremely toxic gas and its presence even at very small concentration in the atmosphere can lead to fatal consequences and, therefore, it affects health, safety, security, and the environment.
- Due to the need to select an appropriate material, it invariably has an economic impact both in terms of materials and fabrication methods.
- The resultant corrosion damage may lead to gradual or catastrophic degradation, all requiring necessary mitigating measures.

H_2S is a weak acid when dissolved in water, hence it affects the solution chemistry in two ways: (i) acting as an acidic gas and hence reducing the in-situ pH, in turn, making hydrogen evolution the primary cathodic reaction; and (ii) catalysing the penetration of atomic hydrogen into the steel which may facilitate hydrogen embrittlement (HE).

The ability of H_2S to influence acidity is indicated by its ionisation, as shown in Eq. (6.1):

$$H_2S \leftrightarrow H^+ + HS^- \tag{6.1}$$

As the H^+ is consumed through a cathodic reaction of hydrogen reduction, more is released and hydrogen gas readily appears on steels exposed to oxygen-free water containing H_2S as shown in Eqs (6.2, 6.3):

$$H^+ + e \rightarrow H\left(\text{atomic hydrogen}\right) \rightarrow H_{ads} \tag{6.2}$$

$$2H_{ads} \rightarrow H_2\left(\text{molecular hydrogen}\right) \tag{6.3}$$

The anion (HS^-) dissociates further to S^{2-} and H^+. The S^{2-} ion reacts with iron to form the black FeS corrosion product commonly found in service.

H_2S or sulphide poisons the reaction Eq. (6.3) and retards the formation of molecular H_2. This increases the surface concentration of H_{ads} and its entry into the steel to cause HE.

6.2.2 Types of H_2S Corrosion Threat

The presence of H_2S and brine in produced fluids not only gives rise to increased corrosion rates, but also can lead to environmental fracture associated with enhanced uptake of hydrogen atoms into the steel.

Wet H_2S primarily causes three main types of corrosion threat as schematically shown in Figure 6.2 including:

- metals loss corrosion – this is dealt with in Chapter 5.
- sulphide stress cracking (SSC).
- stepwise cracking (SWC):
  - blistering
  - hydrogen-induced cracking (HIC)
  - stress-oriented hydrogen-induced cracking (SOHIC).

Since the combination of pH and H_2S partial pressure is related to the intensity of hydrogen charging in ferritic and martensitic steels, these parameters can also be used to help define the likelihood of SSC, HIC/SWC and SOHIC particularly for carbon and low alloy steels (CLASs).

The different types of EC threats damage caused by the presence of H_2S are described in Section 6.2.3.

6.2.3 Categories, Types, Manifestation, and Mitigation Measures of H_2S EC Threats

In-service, hydrogen damage arising from exposure to wet hydrogen sulphide (H_2S) has wide-ranging implications on the integrity of materials used in the industry. The damage falls into four principal categories: (i) HE effects; (ii) hydrogen internal pressure effects; (iii) stress corrosion cracking (SCC); and (iv) damage in related environments such as those containing chlorides, cyanides, etc. These, together with the degree of susceptibility and prevention methods are described in this section and summarised in Table 6.2 with each description elaborated in the following sections.

• **Metal loss corrosion**

• **Sulphide stress cracking (SSC)**

    o **Normally applicable to downhole steels**

        o *(high strength)*

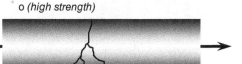

• **Stepwise cracking (SWC)**

    o **Normally applicable to linepipe steel**

        o *(medium to low strength steels)*

1. Blistering
2. SWC
3. HIC
4. SOHIC

Figure 6.2 Principal types of corrosion threat in H_2S-containing environments.

Table 6.2 Categories, types, manifestation, and mitigation measures of EC threats in the presence of H_2S.[a]

Category	Mechanism	Susceptible metals	Key environmental parameters	Causes and manifestation	Manifestation/susceptible locations	Prevention methods
Sulphide stress cracking (SSC)	Hydrogen embrittlement requiring a combination of stress (residual or applied), corrosion process (water and H_2S) and susceptible steel	CLASs, martensitic and duplex stainless steels. (For CRAs, see ISO 15156 Part 3 Table B.1 — Cracking mechanisms that shall be considered for CRA and other alloy groups)	In-situ pH, H_2S partial pressure and temperature (in addition, S^0, chlorides for stainless steels needs to be considered) Refer to ISO 15156	Hydrogen originating from cathodic corrosion reactions behaves as a detrimental constituent.	Can be fast (hours) and catastrophic. Areas of high hardness/high strength/high stress intensity	Control alloy hardness or strength, use of appropriate alloy and, in some cases, stress relieve by heat treatment. Cannot be managed by monitoring.
Hydrogen pressure damage	Cracking of hydrogen embrittled matrix under stress produced by local accumulation(s) of molecular hydrogen. (ISO 15156: HIC; planar cracking that occurs in CLASs when atomic hydrogen diffuses into the steel and then combines to form molecular hydrogen at trap sites)	Susceptible CLASs, flat products (See ISO 15156 Part 2 including: For HIC, see Clause 8: flat rolled CLAS) HIC rare in forgings, castings, seamless pipe		Hydrogen uptake and diffusion result in HIC, SWC and SOHIC	Damage can be seen in various forms depending upon type and location of the inclusions/segregations present and the stress pattern.	Select a high quality clean material and, in some cases particularly for SOHIC, to reduce stresses by heat treatment. In some circumstances appropriate inhibitors can be used. Alternatively, CLAS internally clad with CRA steel may be specified.

Stress Corrosion cracking (SCC)	Crack initiation and propagation by localised (anodic) corrosion. ISO 15156: SCC: cracking of metal involving anodic processes of localised corrosion and tensile stress (residual and/or applied) in the presence of water and H_2S	CRAs, i.e. stainless steels and Ni-alloys (For CRAs see ISO 15156 Part 3 Table B.1 – cracking mechanisms that shall be considered for CRA and other alloy groups)	ISO 15156 Part 3 variables	Through localised corrosion and particularly in the presence of chloride occurring on passive materials when mechanically depassivated by a residual micro-creep	Highly stressed areas and microstructural anomalies	Reducing stress, use of alternative steel or change of environment
Cracking in Related Environments	Not covered in this Publication					

[a] It should be noted the main reference for these EC threats is that of ISO 15156.

6.2.3.1 Sulphide Stress Cracking (SSC)

SSC is a form of HE phenomenon, in which cracking is caused by the dissolution and diffusion of hydrogen into the steel when subject to tensile stress. The primary concern with the presence of H_2S is the risk of cracking where the occurrence primarily affects CLASs and, to a lesser extent, CRAs. The risk of cracking in sour service conditions has long been recognised and has prompted the development of the ISO15156 standard. ISO15156 also referenced as ANSI/NACE MR0175/ISO 15156 is primarily concerned with SSC, although some statements on other types of damage are covered.

The cracking caused by H_2S can result in catastrophic failure in certain circumstances and is often difficult to detect early and monitor in practice, and to subsequently control once initiated. Thus emphasis is firmly placed on identifying the risk at the materials selection stage and to select a material which is not susceptible to cracking. Trying to chemically control the situation subsequently by limiting the level of exposure to H_2S by, for example, treatment with an H_2S scavenger, should be considered a short-term interim response at best and only for relatively low levels of H_2S, not exceeding the general 0.35 kPa (0.05 psi or 0.0034 bar) H_2S partial pressure sour service threshold. Treatment alone with a corrosion inhibitor should be viewed as a high risk option and should be avoided. Encountering during operation unexpected levels of H_2S above that predicted at design, and certainly >0.35 kPa H_2S, should immediately initiate a detailed risk assessment to determine the most appropriate course of action to take.

The atomic hydrogen under the concentration gradient developed through the surface corrosion reaction diffuses into the metal matrix where it can get trapped in inclusions, cavities, and grain boundaries as well as pass right through the metal. It is the higher strength (potentially >70 ksi/480 MPa yield strength) and hardness steels which are susceptible to SSC. This results in localised embrittlement caused by the trapped hydrogen atoms which, once a crack has initiated, will concentrate just ahead of the crack tip and promote its propagation through the metal. Crack initiation is not dependent on pit formation but can occur at any surface stress raiser or discontinuity in the presence of an applied stress. As with all localised corrosion processes, there is an induction period before a crack initiates which will depend on the stress level, local hardness, the material composition/microstructure, and the hydrogen permeation rate. At low stress levels, cracks will tend to be intergranular, whereas at high stress levels, they can be transgranular.

As SSC is associated with higher strength materials, it is generally of greater concern to downhole and topside equipment. In terms of CLAS pipeline operations, SSC is always related to heated affected zones or weld areas with high hardness. Thus, ISO 15156 states that for CLAS weldments to be resistant to SSC, they should generally have a hardness less than HRC 22 and receive where required, appropriate post-weld heat treatment.

6.2.3.2 Hydrogen Internal Pressure Effects

Atomic hydrogen diffuses into the material and recombines as gaseous hydrogen and collects at inclusions, stringers, or other microstructural inhomogeneity where it produces an internal pressure causing another type of EC. This type of threat primarily affects CLAS components and particularly flat products.

This form of hydrogen internal pressure damage for CLAS only illustrated in Figure 6.2 is as follows:

- *Hydrogen blistering:* Hydrogen blistering occurs where inclusions or voids are present in the metal. Atomic hydrogen can diffuse to these locations and convert to molecular hydrogen. Since molecular hydrogen cannot diffuse, the concentration and pressure of hydrogen gas within the voids increase and may be sufficient to cause yielding in the metal and produce a bulge. These voids or inclusions are generally associated with non-metallic inclusions.

- *HIC/SWC:* HIC/SWC is formed in steels by the propagation and linking up of small and moderate-sized laminar cracks in a step-like manner. As more hydrogen diffuses into the steel, the areas around these laminar cracks become highly strained and this can cause linking of the adjacent cracks to form HIC/SWC in the through thickness direction between the individual planar cracks.

  SWC or HIC tends to be associated with lower strength (potentially steels having <80 ksi/550 MPa yield strength) and low hardness CLASs. However, unlike SSC, it does not require the presence of an applied stress. Here the hydrogen atoms entering the metal matrix are able to combine at voids or inclusions within the metal matrix to form hydrogen gas which then builds up internal pressure. Eventually the local material yield strength is exceeded and cracks start to grow parallel to the metal surface with very sharp crack fronts. Many internal cracks can be initiated at any one time which will grow independently without really undermining the integrity of the steel. However, at some stage, inter-crack communication occurs in a stepwise fashion, leading to through wall thickness crack propagation which can then result in ultimate failure. The mechanism which leads to the stepped jump between parallel cracks is a matter of debate – it could be SSC-induced or maybe just a result of the high strain induced around approaching cracks.

  With the new generation of line pipe steels, SWC/HIC is far less of a problem and now rarely encountered. These include much cleaner (ultra-low sulphur content) and refined microstructures (calcium treatment to modify their sulphide inclusion shape into round globular inclusions rather than elongated due to rolling) now produced (i.e. by the thermo mechanical controlled process). However, even where stepwise cracks have formed in the body of a pipe, this may still not seriously undermine integrity in practice; but again should be subject to detail risk assessment once the presence of HIC/SWC is detected.

- *SOHIC (stress-oriented HIC):* In some cases, when metal is subject to stress, small laminar HIC cracks become lined up in the through-thickness direction and step cracks form between them, hence the occurrence of SOHIC. SOHIC is defined as staggered small cracks which are formed approximately parallel to the principal stress resulting in a 'ladder-like' crack array, linking pre-existing HIC/SWC. Formation of this type of damage is linked to particular locations which are susceptible to laminar cracking and to the stress pattern. This is often found, though not exclusively so, in weld heat-affected zones.

  SOHIC, a 'mutation' of SWC/HIC, is a phenomenon resulting from a combination of two independent forms of hydrogen damage: HIC and SSC. New generations of line pipe steel are becoming available offering superior metallurgy with improved strength. These materials, reported to be resistant to either HIC or SSC, have been found to suffer from SOHIC in certain environments. In these circumstances, hydrogen concentration within the lattice is not sufficient to cause conventional HIC, but enough to cause the combination of HIC/SSC in the presence of external stress, hence the occurrence of SOHIC. SOHIC appears on the increase causing a growing concern to address during line pipe steel manufacture and welding qualification acceptance testing. Here the initially formed cracks parallel to the steel surface are stacked one above each other in line through the wall thickness at a region of higher localised stress in the metal matrix – e.g. the edge of the heat-affected zone (HAZ) of a weld. This situation can seriously undermine integrity; and, unfortunately, only appears to occur in the cleaner grades of pipeline steels now produced; and this may also be associated with mixing different sources of nominally the same specified line pipe during field make-up and laying. This phenomenon is still being actively researched to get a better understanding of the mechanism, but susceptibility does appear to be strongly related to the (local) inhomogeneity of the steel.

6.2.3.3 Chloride Stress Corrosion Cracking (Cl⁻SCC)

Cl-SCC is a form of localised corrosion, occurring on passive materials (such as stainless steels or CRAs) when mechanically depassivated by a residual micro-creep. This form is an extension of the 'classical' SCC of stainless steels in aerated brines primarily occurring at temperatures in excess of around 60–80 °C. It can also occur in de-aerated brines when sulphides are present. CLASs are normally immune to Cl⁻SCC and it primarily affects CRAs as outlined in Table 6.1. Cl⁻SCC is characterised by cracks propagating either transgranularly or intergranularly. ISO 15156 Part 3, described in Section 6.4.3, can generally be used as a guide to show the limits of application of CRAs and their resistance to Cl⁻SCC. In general terms, the propensity to Cl⁻SCC increases with increasing temperature and chloride concentration.

6.2.3.4 Cracking in Related Environments

This type of damage includes hydrogen damage or SCC in sour environments in the presence of chlorides, cyanides, alkalis, and amines. This type of damage normally occurs in downstream applications and is not covered in this chapter.

6.2.3.5 Operating Temperatures

Whatever the cracking mode, the rate of hydrogen entry and permeation are important parameters. Increasing the temperature will increase the mobility of the hydrogen atoms. This not only gives rise to a hydrogen recombination reaction but also tends to cause the hydrogen to pass right through the metal without getting trapped in the metal matrix.[1] Therefore, increasing temperature reduces the propensity to SSC. Thus, for example, a steel tubular may exhibit no cracking when in service downhole at a high operating temperature (e.g. 100 °C) but may crack if the well is shut in and allowed to cool or after the tubing has been pulled. A corrosion inhibitor may reduce the rate and amount of hydrogen uptake but what constitutes a safe level is often very difficult to define and, as a general approach, must be viewed as a dangerous strategy, certainly in the long term.

6.3 Current Industry Practices

For the purpose of materials selection and as a description of metallurgical requirements for sour service applications, ISO 15156 has been developed, incorporating the majority of standards and best practices related to the topic. ISO 15156 describes general principles and gives requirements and recommendations for the selection and qualification of metallic materials for service in equipment used in oil and gas production and in natural gas sweetening plants in H_2S-containing environments. It supplements, but does not replace, the material requirements given in the appropriate design codes, standards and regulations.

6.4 ISO 15156

ISO 15156 Standard, again sometimes referenced as ANSI/NACE MR0175/ISO 15156, was developed over the years combining past experience of NACE as reflected in NACE MR0175 while incorporating the current understanding of HE and EC mechanisms and significance of H_2S as captured in EFC Publications 16 and 17. The Standard is in three Parts, summarised in the following for their notable statements.

6.4.1 Part 1

Part 1 of ISO 15156 covers the general principles for the selection of cracking-resistant materials. It deals with the glossary and addresses all the mechanisms of cracking that can be caused by H_2S, including SSC, HIC and SWC, SOHIC, SZC and GHSC.
Part 1 defines:

- Responsibilities of the user and exchange of information.
- Evaluation of service conditions to enable materials qualification for a particular application and/or selection using Parts 2 and 3.
- Materials description and sampling.
- Qualification by field experience or laboratory testing.
- Reporting of qualification and/or selection.

6.4.2 Part 2

Part 2 of ISO 15156 covers cracking-resistant CLASs, and the use of cast irons. It addresses the resistance of these steels to damage that may be caused by SSC and the related phenomena of SOHIC and SZC. It also addresses the resistance of these steels to HIC and its possible development into HIC/SWC. This Part is only concerned with cracking. Loss of material by general (mass loss) or localised corrosion is not addressed.
Part 2 stipulates that SSC Qualification shall require one or more of the following:

- SSC Testing in accordance with the materials Manufacturing specification.
- Testing for specific applications.

- Testing for SSC Regions 1 or 2 (as described in Figure 6.3 and Table 6.3).
- Testing in all SSC regions in Figure 6.3.

The qualification and selection of CLASs with resistance to SSC, SOHIC and SZC are set out in two options taking advantage of the parameters in Figure 6.3 as follows:

- *Option 1* deals with the selection of SSC-resistant steels for conditions where there is no specific information available on the solution chemistry. This Option places a limit of 0.35 kPa (0.05 psi) on H_2S partial pressure – in these conditions, two routes are recommended according to H_2S partial pressure:

  - $pH_2S < 0.35$ kPa (0.05 psi) under which conditions, normally, there are no special precautions required for the selection of steels, nevertheless, highly susceptible steels can crack.[2]
  - For $pH_2S > 0.35$ kPa (0.05 psi) in which SSC-resistant steels in accordance with the requirements of the standard shall be selected.

- *Option 2*: This option allows the user to qualify and select materials for SSC resistance for specific sour service applications or for ranges of sour service where the solution chemistry is known giving flexibility of choice. It facilitates the purchase of bulk materials with economy in mind, although it requires knowledge of both the in-situ pH and the H_2S partial pressure and their variations with time. This is further described in a later section.

Part 2 also covers the requirements for SWC/HIC and acceptance criteria. When evaluating flat-rolled CLAS products for sour service environments, the possibility of HIC and SWC should be considered even in the presence of trace amounts of H_2S. Consideration should also be given to the likelihood of SOHIC and SZC.

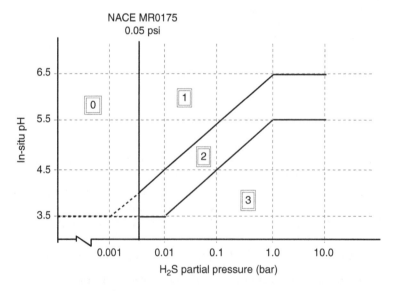

Figure 6.3 Regions of environmental severity with respect to SSC of CLASs as per ISO 15156.

6.4.2.1 Severity of Operating Conditions for CLASs

The severity of sour service condition for CLASs described in ISO15156 is shown in Figure 6.3 in which four SSC regions are identified on the graphical presentation of H_2S partial pressure versus in-situ pH, the two predominant aqueous phase parameters which influence materials performance in sour media. In the context of these domains, decreasing pH or increasing H_2S partial pressure enhances the severity of the damage in sour service.

Regions of sour service shown in Figure 6.3 characterise materials' suitability for sour service applications as outlined in Table 6.3.

It should be noted that some users stipulate that at pH <3.5, the lack of a lower limit for pH_2S means that any detectable trace of H_2S leads to the restriction of Region 3.

6.4.2.2 Key Governing Criteria

For CLASs, the susceptibility to SSC can be characterised in terms of steel manufacturing procedure as outlined below:

1) Seamless tubular steels for well completions: by grade.
2) Flat product steels for pipeline applications: by hardness.

Table 6.3 Typical examples of CLASs for different regions of Figure 6.3.

SSC region	Definition	Metallurgical requirements	Examples	
			ISO 11960/API 5CT	ISO 3183/API 5 L
0	Trace H_2S service	No specific metallurgical precautions are needed	This domain applies to typical CLAS components used for oilfield duties in accordance with construction codes (VHN ≤ 350)	
1	Mild sour service	Minor and inexpensive precautions are required	CLAS tubing and casing up to ISO 11960/API 5CT grade P110 with restricted yield strength or any CLAS of similar resistance to SSC	Homogeneous steels or API 5 L welded pipelines with hardness not exceeding 300 VHN (30 HRC) or any CLAS of similar resistance to SSC
2	Intermediate sour service	Increasing metallurgical precautions are required	Steel tubing and casing up to the API 5CT grade N80 or any CLAS of similar resistance to SSC	Homogeneous steels or API 5 L welded pipelines with hardness not exceeding 280 VHN (27 HRC) or any CLAS of similar resistance to SSC
3	Severe sour service	The most stringent precautions are necessary	CLASs taken from the ISO15156 reference list	Maximum of hardness restriction 250VHN (21 HRC)

These are important considerations in defining the limits of the application of CLASs for sour service applications as detailed earlier.

6.4.3 Part 3

This Part covers cracking resistance and the limitations that apply to CRAs and other alloys appropriately grouped, although some individual alloys from NACE MR0175 (2003 edition which will not be replaced) that did not fit into a group are also included. This Part addresses the resistance of CRAs to damage that may be caused by SSC, SCC, and GHSC. This Part of ISO 15156 is only concerned with cracking. Loss of material by general (metal loss) or localised corrosion is not addressed.

It should be noted that CRAs may undergo different types of cracking subject to the microstructure and operating temperature.

While Figure 6.3 which is devised for CLAS components only can be used as a guide for CRAs in Part 3, it is not directly applicable to CRAs. Limits of application of different categories of alloys are characterised in terms of five operating parameters: maximum allowable temperature, maximum allowable H_2S partial pressure, maximum chloride content, minimum in-situ pH, and presence/absence of elemental sulphur. The materials are then further characterised by their hardness limitation and grade in terms of yield strength.

6.5 Summary

A combination of metallurgical, mechanical, and environmental parameters can unite in causing EC threat to steels. While in the majority of cases, these are specific to an alloy/environment system, their potential occurrence has wide-ranging implications for the integrity of facilities in hydrocarbon production with associated risk and cost penalties. Three principal types of threat driven by H_2S, chloride, or a combination of these are described. ISO 15156 defines the limits of application of different types of steel and materials groups and is a key publication necessary for the design of facilities and equipment. In general, suitability for use in sour service can be characterised by yield strength of well tubular and weld hardness of line pipe steels.

Notes

1 It should be noted that hydrogen damage can also occur at elevated temperature conditions typical of downstream facilities, such as petrochemical or refineries or other industry sectors. However, these are beyond the scope of this book and are not addressed here.

2 Depending on the metallurgical characteristics and alloy chemistry, very high strength steels can be susceptible to SSC in all regions of Figure 6.3 and their use should be made with caution.

Bibliography

Boucher, C, Maltrud, F, Kermani, MB, et al., Oil and gas production: an improved methods for weld sulfide stress cracking resistance assessment. Paper presented at Industrial Heat Transfer Conference, Dubai, 24–26 September, 2000.

Crolet, J.L. (2016). Analysis of the various processes downstream cathodic hydrogen charging, I: Diffusion, laboratory permeation and measurement of hydrogen content and diffusion coefficient. *Matériaux & Techniques* 104 (205).

Crolet, J.L. (2016). Analysis of the various processes downstream cathodic hydrogen charging, II: Charging transients, precharging and natural permeation. *Matériaux & Techniques,* 104 (206).

Crolet, J.L. (2016). Analysis of the various processes downstream cathodic hydrogen charging, III: Mechanistic issues on charging, degassing and sulphide stress cracking. *Matériaux & Techniques* 104 (302).

Crolet, J.L. (2016). Analysis of the various processes downstream cathodic hydrogen charging, IV: Detailed mechanism of sulphide stress cracking. *Matériaux & Techniques* 104 (303).

Institute of Materials (1995). *Guidelines on Materials Requirements for Carbon and Low Alloy Steels for H_2S-containing Environments in Oil and Gas Production*, Publication No. 16. European Federation of Corrosion.

Institute of Materials (1996). *Corrosion Resistant Alloys for Oil and Gas Production: Guidance on General Requirements and Test Methods for H_2S-service*. Publication No. 17. European Federation of Corrosion.

ISO, Materials for use in H_2S-containing environments in oil and gas production, ISO 15156/NACEMR0175, 2015.

Kermani, M.B. (1994). Hydrogen cracking and its mitigation in the petroleum industry. In: *Proceedings of the Conference on Hydrogen Transport and Cracking in Metals* (ed. A. Turnbull), 1–8. London: The Institute of Materials.

Kermani, MB, Materials optimisation for oil and gas sour production, NACE Annual Corrosion Conference, Paper No. 00156, 2000.

Kermani, MB, Boucher, C, Crolet, JL, et al., Limits of weld hardness for domains of sour service in oil and gas production. NACE Annual Corrosion Conference, Paper No. 157, 2000.

Kermani, MB, Gonzales, JC, Turconi, GL, et al., Materials optimisation in hydrocarbon production. NACE Annual Corrosion Conference, Paper No. 05111, 2005.

Kermani, MB, Harrop, D, Truchon, MLR, and Crolet, JL, Experimental limits of sour service for tubular steels. NACE Annual Conference, CORROSION '91, Paper No. 21, Cincinnati, Ohio, March 1991.

Kermani, M.B. and Morshed, A. (2003). CO_2 corrosion in oil and gas production: a compendium. *Corrosion* 59: 659–683.

7

Corrosion in Injection Systems

Reservoir management of hydrocarbon resources and increased production are among the key challenges facing the hydrocarbon production industry sector. These invariably necessitate the use of water and/or gas injection to maintain or increase reservoir pressure in response to a decline in natural lift and to enhance recovery as well as emission control, to minimise the impact on the environment. Water injection in particular is increasingly used for this purpose.

Here the use of carbon and low alloy steels (CLASs) for injection tubing and flowlines is the preferred option for reasons of economy and availability. The primary corrosion threat in such systems is by dissolved oxygen which may be present in the water intended for injection. Dissolved oxygen is commonly controlled first by physical removal – e.g. gas stripping, vacuum deaeration or membrane contactors – followed by final treatment down to typically <15 parts per billion (ppb) by continuous injection of an oxygen scavenger. It is not usual practice to treat injection water with a corrosion inhibitor alone or in addition to deoxygenation, although there can be compatibility issues with other chemicals including oxygen scavengers (OS) that need to be taken on board.

Water is also deployed in hydrotest packages, and Section 7.8 has been assigned to briefly address this topic.

This chapter focuses on the key elements affecting corrosion threats in water injection systems, outlining the means of prediction, mitigation, and control. It takes advantage of industry-best practices with a view to providing parameters and measures within which a trouble-free operation can be achieved with safety in mind. In addition, a brief reference to water treatment methods is given. The chapter focuses on the use of CLASs. In addition, materials options for injection applications are briefly covered.

Another form of corrosion threat in water injection system is related to microbial activity, which is the subject dealt with in Chapter 11.

The present chapter covers four sections including:

1) an overview of injection systems;
2) water corrosivity, types of corrosion and means of prediction;
3) types of injection systems and respective corrosion threats;
4) choice of materials for injection duties.

Corrosion and Materials in Hydrocarbon Production: A Compendium of Operational and Engineering Aspects, First Edition. Bijan Kermani and Don Harrop.
© 2019 John Wiley & Sons Ltd. This Work is a co-publication between John Wiley & Sons Ltd and ASME Press.

7.1 The Intent

Injection has proved to be one of the most economical methods for reservoir management. The technology is invaluable in helping maintain reservoir pressure, enhancing the production of hydrocarbon reserves, and reducing the environmental impact through re-injection of treated and filtered produced (reservoir/formation) water, other waters, and/or gas. To achieve this, pipelines and wells have been designed according to the type of fluid/gas intended for injection.

7.2 Injection Systems

Different types of injection systems are used as summarised in Table 7.1, which identifies the respective corrosion threats and corrosion prediction tools. Injector systems are system-specific, and the corrosion associated with each result from the associated gas

Table 7.1 Injection types and primary cause of corrosion

Injection type	Cause of corrosion	Corrosion prediction tool	Remedial measures	Complementary measures
Produced water (PW)	Primarily CO_2/H_2S	Refer to CO_2 corrosion models	Acid gas removal, and/ or inhibition	—
Sea water, brackish, aquifer or river water	Primarily O_2	Covered in this chapter	O_2 removal	Biocide treatment of sea water at least with continuous injection of chlorine (or hypochlorite) as a biocide often supported with regular batch treatment – typically weekly with an organic biocide
Sea water/PW (commingled water)	Both O_2 and CO_2/H_2S	No models available	Acid gas and O_2 removal	
Water alternating gas (WAG)	Both O_2 and CO_2/H_2S	A combination of CO_2 corrosion models and those covered in the present chapter. N.B. the actual unmitigated corrosion rates can be synergistic in magnitude and localised in nature.	Acid gas and O_2 removal	
Gas injection (gas lift or CO_2 sequestration)	Primarily CO_2/H_2S (if not dried)	Refer to CO_2 corrosion models	Drying or inhibition (no realistic model for CO_2 injection for sequestration)	—

is outlined in Table 7.1. NACE TM0299 [1] also provides the necessary corrosion control measures for water injection systems. This section describes these further.

CLASs are the preferred materials of choice, subject to meeting the necessary mechanical properties and corrosion performance. In the selection of an appropriate material, a corrosivity assessment is carried out in respective media as outlined in Table 7.1. When the primary corrosive agent is CO_2 (or H_2S), corrosion prediction is performed using an available model, such as those described in Chapter 5. However, in the presence of O_2, other models are available as described in Section 7.3.

7.2.1 Treated Water

Water treatment is intended to produce and deliver water of a given quality to the injection wellbore in order to do the following:

- achieve a given design life;
- minimise the generation of suspended solids;
- avoid reservoir souring;
- remove any risk of contaminating the open environment in which the operating facilities are located.

These objectives are met through delivering water by a combination of corrosion mitigation measures captured in Figure 7.1, including:

- chemical control of water by the use of corrosion inhibitors, biocides, OS;
- materials choice through appropriate selection of CLAS, CRA, non-metallic;
- internal coatings and linings by the use of high/medium density polyethylene (H/MDPE), internally plastic coated tubular (PCT), internally lined glass reinforced epoxy (GRE) CLAS;

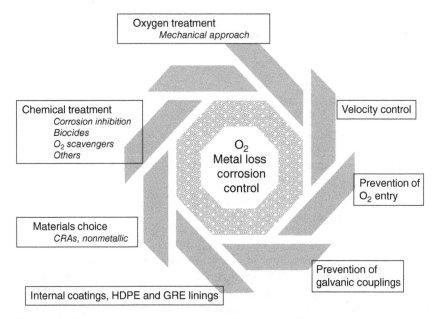

Figure 7.1 Oxygen corrosion control measures in water injection systems.

- oxygen removal by mechanical approach;
- velocity control using appropriate tubing size, taking account of API RP14E guideline;
- prevention of oxygen entry and couplings of dissimilar metallic materials.

It should be noted that depending on the injection system and the respective media, the choice of gas removal depends on environmental restrictions and economic constraints, otherwise all gases can be removed – H_2S is an exception due to its toxicity.

A number of points are worthy of note here including:

- *Biocide treatment:* this is essential to minimise the possibility of microbial activities: in such circumstances, chlorine and an organic biocide treatment are normally used. Continuous treatment with nitrate to suppress sulphate-reducing bacteria (SRB) activity has grown in usage, although its effectiveness appears to have been variable.
- *Oxygen scavenger:* there is a need to exercise great care in trying to solely treat with an OS as over-dosing can itself cause corrosion and internal fouling (FeS) and is a costly option to scavenging from part per million (ppm) down to part per billion (ppb) level. Furthermore, the system may not have the injection capacity available to carry it out – this is only normally considered in an emergency situation for once through systems.
- *Mechanical treatment:* the sole use of mechanical methods of oxygen removal will not get the dissolved O_2 level down to the required ppb level. This is usually achieved by replacing dissolved O_2 by another soluble gas such as CH_4 or by vacuum de-aeration (reduce the partial pressure of O_2 in the gas, thus releasing dissolved O_2). There is a growing use of reverse osmosis, particularly where there is a desire to remove dissolved ions such as sulphate to stop $BaSO_4$ formation in the reservoir.

7.3 Water Treatment Methods

Successful use of CLAS requires that injected water should undergo a series of treatments. The main physical treatments of injection water are:

- filtration
- de-oxygenation.

Filtration is performed to remove particulates, macro-invertebrates, and organic matter down to less than $5\,\mu m$ in size. De-oxygenation is intended to remove the bulk of dissolved oxygen from the injected water to allow more effective use of CLAS and reduce aerobic bacterial growth. This is carried out either by: (i) mechanical or (ii) chemical methods or a combination of these. For gas transmission lines, dehydration is good practice, ensuring no condensation.

7.3.1 Mechanical Treatment

Mechanical methods include: (i) gas stripping using an inert gas or possibly associated natural gas; or (ii) vacuum deaeration systems. Through these methods, oxygen reduction from 8 ppm dissolved in fully saturated water to as low as 10 ppb can be achieved if fully and effectively operational, although, as discussed above, additional chemical treatment is required to achieve such low levels. Further oxygen removal, if required, can be achieved by the addition of an oxygen scavenger downstream of the deaeration vessel (chemical treatment). It is vital that low oxygen concentrations are achieved to protect all CLAS components, including topsides equipment, injection wells, and subsea flowlines from O_2 corrosion.

The important messages in oxygen removal and corrosion control are:

- A distinction should be made when assessing potential corrosivity between gas stripping systems and vacuum deaerated systems because of the likely synergy between CO_2 and O_2. Higher corrosion rates are expected in gas-stripped systems for the same O_2 concentration due to the presence of CO_2 when using natural gas.
- Corrosivity should be assessed on a system-by-system basis. Only then may changes to the operating parameters (e.g. increased oxygen or chlorine limit) be considered.

7.3.2 Chemical Treatments

The main areas for the application of chemical treatment associated with water injection systems include the following.

7.3.2.1 Oxygen Removal

Oxygen removal by chemical treatment is carried out using an OS such as hydrazine or ammonium or sodium bisulphite/sulphite where the latter are often catalysed to speed up their reactivity with dissolved O_2. Due to cost implications, chemical oxygen removal alone is only carried out during upset conditions or when mechanical removal is not operational. The use of hydrazine is very uncommon, if at all, as it is a very hazardous chemical to handle.

7.3.2.2 Injectivity Problems and Drag Reduction

In order to maximise the injectivity of water into a reservoir and to increase efficiency in a sweep of reservoir fluids, a polymer flow improver/drag reducer can be applied. This is particularly applicable to situations where no spare injection capacity exists, or it is not logistically or economically viable to introduce – i.e. not feasible to drill further water injection wells.

7.3.2.3 Coagulants and Filter Aids

These chemicals are applied to improve the efficiency of the filtration system and aid the removal of organic components (e.g. oil droplets in re-injected produced water), which can contribute to foaming of injection water. Coagulants and filter aids typically work at very low dose rates.

7.3.2.4 Bacterial Growth and Proliferation

Biocide is injected into the bottom of the raw sea water lift pump casing to control marine fouling and bacterial growth and minimise biological activities, particularly in long lines and under elevated temperature conditions before entering the injection water treatment facilities. To then achieve the desired injection water quality, bacterial control conducted alongside oxygen removal includes treatment with:

- chlorine or hypochlorite solution (OCl^-);
- organic biocide, e.g. glutaraldehyde;
- ultraviolet rays.

Bacterial control is achieved by continuous chlorination treatment upstream of deaeration together with periodic batch biocide addition – as a shock treatment – downstream of deaeration. Depending on which biocide is used, the biocide can interfere with the oxygen scavenger and therefore the scavenger may need to be turned off during batch treatments.

7.3.2.5 Antifoam

Many oils foam when trapped gas is suddenly released under conditions of an abrupt drop in pressure. This is undesirable, causing process and transportation challenges. Therefore, antifoams are required, in which case they are generally applied upstream of de-aeration if required – usually where vacuum deaeration is being used. These are typically silicone-based fluids or polyglycols.

7.4 Water Corrosivity

Bearing in mind the other types of corrosion threats shown in Table 7.1, corrosion in water injection systems is primarily due to dissolved oxygen content. An appropriate choice of materials for injection application is therefore subject to the available water quality and its continual maintenance.

The principal types of corrosion experienced in injectors are summarised in Table 7.1 and fall into two broad types:

- CO_2/H_2S corrosion when injecting:
  - produced water (PW), commingled water (PW mixed with another source such as sea water), produced gas, water alternating gas (WAG), etc.
- O_2 corrosion when injecting:
  - sea water, brackish, aquifer or river waters, WAG, commingled water.

There are several models available to predict the corrosion of CLAS in the presence of oxygen as affected by operating conditions. These use a mass transfer expression as explained in the following sections.

It has been recognised that high oxygen excursions, even for short periods, must be avoided. This is not only because of the effect of high oxygen levels on the corrosion of CLAS but also because protective scale breakdown is more likely which could lead to enhanced localised attack.

Finally, dissolved oxygen can be fairly quickly consumed by the reaction with carbon steel piping but requires longer lengths of piping than that typically available on offshore/topside facilities. Nevertheless, the length of piping between de-aeration and injectors may reduce the potential corrosivity to some degree. However, the use of CRA piping/flowlines, if economically viable, and the increasing use of plastic-lined flowlines not only benefit corrosion resistance but also reduce liquid drag at the pipe-wall. However, this would transfer all the risk of oxygen corrosion, due to residual O_2 levels and during upsets, downhole. Under such circumstances the use of plastic or GRE-lined injection tubing should also be considered.

7.4.1 Water Quality

Water quality is paramount in affording low corrosion of CLAS components. Water quality is characterised by: (i) oxygen content; (ii) residual biocide; and (iii) the presence of solids. These, together with fluid velocity, affect the corrosion performance of materials, an overview of which is provided in this chapter. The following captures limits imposed on water quality and conditions to allow successful use of CLAS – they fall into four categories:

- *Oxygen content:* removal of residual oxygen in the water prior to injection and making certain that oxygen is maintained below 20 ppb (and typically targeted at ≤10 ppb). It is important to note that intermittent spikes of 50–100 ppb can over time result in localised corrosion becoming established, influenced by local flow conditions and poor solids control.
- *Solids content:* injected water should be nominally free from solids (<1 lb/1000 bbl).
- *Residual Cl$_2$ or biocide content:* injected water should contain zero residual chlorine (maximum 0.3 ppm) in the system – an organic biocide (glutaraldehyde) treatment is the preferred choice when using CRAs. A stringent biocide programme from the beginning of water injection is strongly recommended, as included in EFC Publication 64.
- *Velocity:* Injected water velocity should be below 8 m/s and/or below the API RP 14E erosion limit with c = 250.[1]

7.5 Means of Corrosion Prediction

There are several models used by the industry to predict corrosion of CLAS in O$_2$-containing media, some of which are described in this section.

7.5.1 Oldfield and Todd

Oldfield and Todd's [2] expression has been widely used in predicting the corrosion of CLAS in oxygen-containing waters in pipe flows. The model was originally developed for application to desalination systems, hence its questionable suitability to water injection systems required to operate at <50 ppb levels continuously. The expression for a laminar flow regime is as follows:

$$CR = 5.65 \, CV/(Re^{0.125} \, Pr^{0.75})$$

where:

V = the velocity $(m\,s)^{-1}$
C = the oxygen concentration (ppb)
Re = the Reynolds number
Pr = the Prandlt number
CR = the corrosion rate $(mm\,y)^{-1}$.

By including the diffusion coefficient of oxygen in sea water as a function of temperature, the expression can be simplified as follows:

$$CR = C/189\left(V^7/d\right)^{1/8} \exp\left(t/42.6\right)$$

where:

d = inside pipe diameter (m)
t = temperature (°C).

This expression provides an over-estimated corrosion rate as it was developed for the desalination industry except at low oxygen concentrations and flow rates. The Oldfield and Todd model is normally applicable to low oxygen levels (~20 ppb) and low flow rates (~1 m s)$^{-1}$. The model was not intended to be applied to conditions of very high oxygen concentrations (>100 ppb).

Over the years and through comparisons made between predicted CR from Oldfield and Todd and field monitoring, it has been shown that Oldfield and Todd over-estimated CR by a factor of 3–20 times. For design purposes, therefore, a 'rule of thumb' correction factor of 1/3–1/5 (or even 1/10), subject to water quality, is applied to the Oldfield and Todd model to give a first pass predicted oxygen corrosion rate. The correction factor is applicable to:

- velocities <10 m s^{-1}
- API RP 14E limit of c = 250 for velocity limit.

7.5.2 Berger and Hau

Again, for oxygen corrosion, the maximum possible corrosion rate (proportional to i_{lim}) can be related to the mass transfer coefficient, k and the oxygen concentration of the solution, C, as follows:

$$i_{lim} = 4FkC$$

Berger and Hau [3] used this in turbulent flow to correlate the mass transfer to fluid flow and the expression of:

$$Sh = 0.023\,Re^{0.8}\,Sc^{0.33}$$

where:

Re = the Reynolds number
Sc = the Schmitt number (dimensionless numbers used in fluid mechanics).

Having considered various constants, based on the Berger and Hau model, the corrosion rate in oxygen-containing water can be summarised as follows:

$$CR = 0.226\,VC/Re^{0.14} \cdot Sc^{0.706}$$

where:

V = the velocity $(ms)^{-1}$
C = the oxygen concentration (ppb)
CR = the corrosion rate $(mmy)^{-1}$.

However, this prediction is based on average flow conditions, and local flow disturbances might give local corrosion rates possibly five or more times higher [3, 4] and care should be exercised when using this model for complicated configurations.

Other available mass transfer expressions include Chilton and Colburn [5], plus Shaw and Hanratty [6]. Comparisons made between these three models and the model of Oldfield and Todd show that the former three are somewhat similar in their prediction of oxygen corrosion and some 10–20 times lower than that predicted by Oldfield and Todd as described by Andijani and Turgoose [4].

7.5.3 The Appropriate Model

The current industry practice is based on the model developed by Berger and Hau [3]. For the localised corrosion, if any, at hot spots or where there are local flow disturbances, an increased corrosion rate of five times is considered for the Berger and Hau model. Apart from this, the industry continues to use Oldfield and Todd as a routine way to predict corrosion of CLAS in water injection systems, as affected by flow dynamics, O_2 content and operating temperature, although a correction factor is normally adopted.

7.6 Materials Options

In water injection systems, upstream of water treatment equipment, corrosion-resistant alloys (CRAs) are used and thereafter all tubing and equipment is normally made of CLAS. This is subject to ensuring and continually maintaining water quality within a certain quality and the operational levels described earlier. Nevertheless, it should be emphasised that corrosion should be managed on a system-by-system basis by employing continuous monitoring of water injection systems to achieve optimum control. Corrosion monitoring is an integral requirement of implementing an effective corrosion mitigation programme, as noted in Section 7.7.

Throughout the industry, as fields become mature and production falls off plateau and water production increases, interest is growing in the possibilities of water flood, in which consideration is given to water disposal and injection of non-treated water. In these cases, other issues become important, including the management of raw water and increasing interest in commingled water injection, some of the topics briefly described earlier. A summary of the type of materials for injectors is shown in Figure 7.2, with cost increasing from CLAS to CRAs with water quality requirements becoming less stringent.

7.6.1 Tubing

7.6.1.1 CLAS
CLAS has proved an effective choice for water injection tubing subject to meeting stringent requirements on water quality described in Section 7.4.1, adherence to which is paramount.

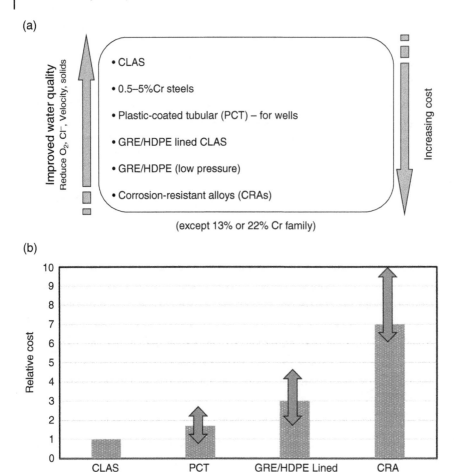

Figure 7.2 A brief list of (a) materials' options; and (b) relative costs for injection application.

7.6.1.2 Low Cr-Containing Steels

Low Cr-containing steels (0.8–5.0%Cr) have been successfully used by the industry, with laboratory results showing slightly improved corrosion performance compared to CLAS, particularly for application in CO_2-containing media, although the outcome is not consistent. The use of low Cr-containing steels cannot therefore provide a fit and forget solution for critical applications or where water quality cannot be maintained. However, they can be considered for wells where water quality is as described earlier or for less 'critical' wells.

7.6.1.3 Plastic-Coated Tubulars

Several internally plastic-coated tubing (PCT) systems are available, most based on liquid-phenolic or powder epoxy novolac compounds. Industry experience of their use is mixed. The attitude to PCT is generally that of a temporary fix rather than a permanent solution to corrosion protection. However, PCT offers drag reduction and other benefits including reduction in/absence of surface scale formation.

The issues related to PCT are its low resistance to wireline damage, well stimulation chemicals, and potential coating collapse in the event of rapid decompression of wells. There is also a risk of possible enhanced crevice corrosion at places where the coating is damaged.

As a result of poor mechanical robustness, the use of PCT in injection service is generally subject to the same corrosion limitations as described for bare CLAS. There are only a limited number of circumstances in which PCT should be considered:

1) When handling non corrosive or adequately and reliably treated fluids.
2) When there is no intervention work or extensive use of acidic chemicals.

However, it should be noted that in corrosive service, PCT will ensure much better final mechanical integrity than plain carbon steel, and hence is likely be easier to retrieve.

7.6.1.4 Glass Reinforced Epoxy-Lined CLAS Tubing

This product consists of a glass reinforced composite/epoxy (GRE) liner which is cemented in place in the internal diameter of standard CLAS tubing. The use of GRE-lined CLAS tubing in wells is becoming more popular within the industry, especially for new projects. However, unless the produced water is suitably cleaned up to minimise the presence of any residual oil, care should be taken when selecting a lining for produced water injection, as most have limited hydrocarbon service capabilities. The points to consider are:

- the top temperature limit of GRE;
- special requirements for joints;
- performance in the presence of well-treating chemicals, although this may be an unlikely scenario for injectors;
- reduced internal diameter due to GRE thickness;
- care should be exercised if wells are to be turned round to producers at a later stage due to potential permeation of hydrocarbon through the lining.

7.6.1.5 CRAs

The use of CRAs may prove beneficial in situations where water quality cannot be maintained satisfactorily, oxygen removal may not be feasible, oxygen excursions may be foreseen, or for untreated water injection. In these situations, only fully passivated alloys are suitable as indicated in Figure 7.2. This excludes families of 13%Cr which have shown poor resistance in these applications, and potentially 22%Cr duplex stainless steel: generally for raw sea water service, the minimum grade of CRA is super duplex stainless steel [7, 8].

In addition, ample application of thread compounds (dope) on couplings should be carried out to minimise crevice corrosion, particularly when using CRAs unless a dope-free coupling is used.

7.6.2 Pipelines and Piping

7.6.2.1 Bare CLAS

The use of CLAS with adequate corrosion allowance has been used for the transportation of water, and its use is subject to meeting stringent requirements on water quality described in Section 7.4.1.

7.6.2.2 Internally Lined CLAS

CLASs internally lined by high/medium density polyethylene (HDPE or MDPE) have been successfully used in industry. This is by means of putting the liner inside the pipeline and positioning a flange connection every kilometre or so. It is apparent that the deployment can be challenging but this has seen a growth in use for new projects especially located in remote or hostile environments.

7.6.2.3 CRAs

Again the use of CRAs may prove beneficial for short lines and in situations where water quality cannot be maintained satisfactorily, oxygen removal may not be feasible, oxygen excursions may be foreseen or for untreated water injection. The conditions are as outlined earlier for tubing applications:

1) potential materials options;
2) approximate relative cost.

7.7 Supplementary Notes

In the selection of an appropriate material for injection application, care should be exercised not only on the corrosion performance under normal operational duty, but also in conditions resulting from other required activities during the life of the system. Here five issues are worthy of particular attention:

- *Acid treatment:* well stimulation requires the use of highly corrosive chemicals – the performance of all tubing materials in this type of treatment, if required, needs to be established. An appropriate well stimulation guideline requires the inclusion of an inhibitor package and strict control of soak time – the optimum inhibitor performance will apply only for a predetermined exposure window – with neutralisation of acid flow-back. This is particularly true when using PCT.
- *Wireline damage:* well logging tools may cause damage to tubulars – this is particularly true for CRAs and PCT, in which case resistance to this type of damage needs to be addressed.
- *Rapid decompression (RD):* the effects of RD must normally be considered in all high pressure gas applications. However, downhole, the effects of RD are most often seen on tool or seal retrieval. Only in shallow gas applications, e.g. shallow set safety valves, is RD ever a real problem downhole. Seal design (seal cross-section and initial compression) and materials selection must be appropriate. The effect of RD is also a major issue when using a PCT or GRE liner, which may lead to coating collapse, albeit to a lesser extent for GRE-lined CLAS.
- *Preferential weld corrosion (PWC):* preferential weld attack is not relevant to well tubing but an area of great importance in water injection system piping and flowlines, in which selection of weld metallurgy is of prime importance – especially the Ni content which typically needs to be present at ~0.7–1% to avoid PWC.
- *Corrosion monitoring:* effective corrosion monitoring in addition to monitoring of water quality in line with what has been discussed earlier are integral and essential requirements – for corrosion monitoring, typically weight loss coupons and electrical-resistant (ER) probes are deployed.

7.8   Hydrotesting

Before new or rehabilitated pressure containing equipment and systems are placed into service, it is normal procedure to test them for integrity at a pressure above their designed maximum working pressure – typically the safety factor is c. 150% of the design maximum working pressure. The testing medium will vary, depending on the ready source and location but may include clean freshwater, potable water, treated sea water, a mixture of water and glycol, oil, air or inert gases; the latter three being more common where testing low-volume equipment. However, by far the most commonly used media is water which may remain in the system, and particularly in pipelines, over an extended period for weeks to a few months is not untypical before being displaced/flushed out.

Water comes from one of several sources including aquifers, rivers, ponds/lakes, or the sea. The type of water used during hydrotesting depends on the water source availability, the volume of water required, and the criticality of the tested equipment and material. For example, for martensitic, austenitic (300 series and Alloy 825), and duplex stainless steels, fresh water is normally specified with a limit placed on chloride content (typically <50 ppm) and pH (6.5–7.5).

Water from any of these sources has an inherent corrosivity, not least due the presence of dissolved oxygen, and the potential to introduce bacteria into a system during hydrotesting. These two types of corrosion threat are dealt with in Chapters 4 and 11, respectively, where hydrotesting severity is dependent on:

- type and quality of water used;
- the length of time water remains in the system;
- the temperature;
- the metallic material exposed to the hydrotest medium.

Corrosion damage may result during the hydrotest period when fully flooded with water or after removal (dewatering and drying), if incomplete or poorly controlled. The corrosion that initiates in this period prior to start-up may continue during the system service life and potentially place added stress on the performance of a material and any required corrosion mitigation being deployed (e.g. corrosion inhibitor). Furthermore, it may be some considerable time into the service life before the damage is first detected, bringing uncertainty over its origin, i.e. during hydrotesting or once in service, and the history/level of continuing activity.

The base case chemical treatment for all sources of hydrotest waters is with an oxygen scavenger and biocide. As most biocides react with an oxygen scavenger, this should be added first, followed by a sufficient lapse of time for it to fully react with all the dissolved oxygen present before the biocide is then added. The concentration of biocide used should factor in the need to account for some loss, due to reaction with the residual (unreacted) oxygen scavenger still present: the concentration of oxygen scavenger will be chosen to give a known excess above that stoichiometrically required.

Treatment of hydrotest water with a corrosion inhibitor is generally less of a critical requirement. Where use may be deemed advantageous is in expected situations of long exposure time (>6 months) to the hydrotest water, noting that any significant chemical depletion in the corrosion inhibitor concentration is far less likely. Nevertheless it is

Table 7.2 Chemicals for hydrotesting water.

Chemical	Period of treatment	Typical dose rate (ppm in fresh or saline water)
Oxygen scavenger (OS)	Added to the hydrotest water	125
Biocide	Short dwell time (<4 weeks)	100–200
	Long dwell time	200–500
O$_2$ corrosion inhibitor	Short dwell time (<4 weeks)	100
	Long dwell time	500
Dye (only for offshore lines)	—	—

not uncommon to see corrosion inhibitor offered as part of a hydrotest chemical treatment package – e.g. it is common to use quaternary ammonium and an amine inhibitor species, which have a degree of tolerance to dissolved oxygen, limited incompatibility with an oxygen scavenger, and have biocidal properties. Corrosion inhibitors are only effective where the hydrotest water is in contact with CLASs. However, in most cases, the addition of a corrosion inhibitor should not be necessary if the selection and deployment of the oxygen scavenger/biocide treatment package have been soundly managed and afforded the necessary importance and priority throughout. Unfortunately, time constraints, schedule, installation and commissioning delays can all too often complicate matters and introduce the risk of compromise and short cuts creeping in.

Table 7.2 gives indicative dose rates for chemicals for treating hydrotest water: dye is added to visually aid leak detection. The chemical service company supplying the treatment package will advise on the actual concentrations required on a system-by-system basis.

Eventually the hydrotest water needs to be drained. This is normally done at the seaward end of offshore pipelines. However, in any location, the treatment chemicals (biocide, oxygen scavenger, corrosion inhibitor, dye) need to adhere to local environmental legislation affecting handling and discharge/disposal. These regulations are becoming increasingly more stringent in many parts of the world. There is a push to develop more environmentally friendly oilfield production chemicals across the spectrum of application. In the case of hydrotest chemicals, biocide has received the most attention with particular focus on biodegradability versus required efficiency and time to remain active above a threshold concentration when deployed. As a result, there is increasing use of the biocide tetrakis (hydroxymethyl) phosphonium sulphate (THPS) offering low environmental toxicity.

Ideally, hydrotest water should be removed as soon as possible from all equipment, facilities, and pipelines exposed, which are then dried internally – e.g. flushed with an inert gas such as dry N$_2$ – ahead of final hook-up and commissioning. The criticality of the drying process will be determined by the corrosion risk associated with any water remaining: here particular attention should be given to identifying dead legs. Vapour phase corrosion inhibitors (VCIs) have also been deployed as part of the drying process with mixed results.

A number of measures are also important to ensure full preparation for the discharge of water, including:

- neutralising chemicals identified, in place, and ready for use;
- neutralising treatment procedure, in place ahead of initiation it use;
- neutralising chemicals selected at the same time as the biocide.

A few additional points in relation to the use of CRA materials worthy of note are as follows:

- Depending on the water type and chemistry, CRAs may be subject to much more severe corrosion attack than CLASs. Hence, the quality of water (source, chemical composition, oxygen content, added chemicals, etc.) used for hydrostatic testing and subsequent immediate and thorough dewatering and drying often are much more critical than in the case of CLASs.
- The corrosion threat to CRAs during hydrotesting is mainly in the form of pitting and crevice corrosion where the effect of biofilm is the reduction of the pitting resistance of stainless steels, although CRAs are not totally immune to MIC.
- The selection of high purity water for hydrotesting may be critical to prevent corrosion in CRA components.
- Weldments have been by far the most MIC-susceptible areas, but corrosion attack also has been observed in the base material.

7.9 Summary

Metal loss corrosion driven by dissolved O_2 in water injection systems remains a key challenge for the successful use of CLAS components. While fully treated water within the limits outlined in this chapter can lead to no corrosion of CLAS, maintaining these limits continually is difficult to implement in operations. Other notable points in this chapter are:

- Water injection is the normal approach to achieve enhanced hydrocarbons recovery and increased production. This can be achieved by injection of several types of fluids and/or gas in which each may result in respective metal loss corrosion threats.
- Several models are available offering metal loss corrosion prediction in O_2-containing systems, each with their own advantages and disadvantages but the Berger and Hau model appears most appropriate and widely used.
- Fully treated injection water achieving O_2 levels below ≤10 ppb results in very low corrosivity to the use of CLAS.
- Some measures to allow effective corrosion mitigation in water injection systems are outlined, highlighting the merits of different materials and that of chemical and mechanical approaches.
- Elements required to treat water for hydrotesting are presented.

Note

1 The c constant has a value specific to materials type as defined in API RP 14E.

References

1 NACE, Corrosion control and monitoring in seawater injection systems. NACE TM0299, 1999.
2 JW Oldfield, GL Swales, and B Todd, Corrosion of metals in deaerated seawater. Paper presented at NACE Middle East Corrosion Conference, Bahrain, 1981.
3 Berger, F.P. and Hau, K.F. (1977). Mass transfer in turbulent pipe flow measured by the electrochemical method. *International Journal of Heat and Mass Transfer* 20,: 1185–1194.
4 I Andijani and S Turgoose, Prediction of oxygen induced corrosion in industrial waters. IWA (International Water Association) Conference, Cranfield, 2003.
5 Chilton, T.H. and Colburn, A.P. (1934.). J-factor analogy. *Industrial and Engineering Chemistry* 26: 1183–1187.
6 Shaw, D.A. and Hanratty, T.J. (1977). Influence of Schmidt number on the fluctuations of turbulent mass transfer to a wall. *AIChE Journal,* 23 (28): 160–169.
7 MB Kermani, JC Gonzales, and GL Turconi, Materials optimisation in hydrocarbon production. NACE Annual Corrosion Conference. Paper No. 05111, 2005.
8 T N Evans, P I Nice, M J Schofield, and K C Waterton, Corrosion behaviour of carbon steel, low alloy steel and CRAS in partially deaerated seawater and comingled produced water. NACE Annual Corrosion Conference, Paper No. 04139, 2004.

Bibliography

ASTM, Standard test method for iron bacteria in water and water-formed deposits. ASTM D932, 1997.
Jones, L.W. (1988). *Corrosion and Water Technology for Petroleum Producers.* Tulsa, OK: OGCI.
Kermani, B. and Chevrot, T. (eds.) (2012). *Recommended Practice for Corrosion Management of Pipelines in Oil and Gas Production and Transportation. The Institute of Materials.* European Federation of Corrosion, Publication No. 64.
NACE, Field monitoring of bacterial growth in oilfield systems, NACE TM0194, 2014.
Patton, CC, Applied Water Technology, 2. JMC, 1995.

8

Corrosion Mitigation by the Use of Inhibitor Chemicals

The most commonly used approach to manage the threat of internal metal loss corrosion of carbon and low alloy steels (CLASs) exposed to produced fluids containing acid gases is treatment with a corrosion inhibitor. For such applications, corrosion inhibitors have been the primary corrosion control option for over 60 years. In the early 1950s the discovery that high molecular weight, long-chain organic molecules behaved as corrosion inhibitors revolutionised the upstream oil and gas industry, providing the ready and flexible ability to effectively treat sweet and sour systems and therein greatly extend the use of CLASs.

A corrosion inhibition system generally requires relatively low capital outlay but carries a continuous whole life operating cost – cf. ongoing costs of chemical; deployment, and injection system management and associated maintenance; supporting corrosion monitoring and inspection. Viewed on a Net Present Value (NPV) basis, corrosion inhibition with the use of CLASs generally proves the most economically attractive option at the design stage of project. These elements are explored briefly in this chapter which again focuses on engineering aspects and refers the reader to other publications for further details.

The most widely used and invariably the most cost-effective and practical method of application is by continuous injection of corrosion inhibitor into the produced fluids. This chapter therefore concentrates primarily on inhibitors used in this manner unless specifically stated otherwise. There is limited discussion here of other methods of application in the field, which are based on batch treatment, used far less commonly, and then almost exclusively for downhole treatment of wells.

8.1 Inhibitor Characteristics

Commercial corrosion inhibitors are formulated products designed to be readily injected and dispersed into a system and transported to the metal surfaces at risk of corrosion where they interact through adsorption to form a protective surface film. A formulated product does not just contain active inhibitor species. Formulation may well be specific to the application conditions being considered in order to achieve optimum performance. Formulation can also differentiate performance between commercial products with the same active inhibitor species, can affect unit

Corrosion and Materials in Hydrocarbon Production: A Compendium of Operational and Engineering Aspects, First Edition. Bijan Kermani and Don Harrop.
© 2019 John Wiley & Sons Ltd. This Work is a co-publication between John Wiley & Sons Ltd and ASME Press.

cost, and is a key factor in the ease and effectiveness of deployment throughout a production system. It is because of the need and importance of formulation that there is not, at least so far, a 'silver bullet' corrosion inhibitor suitable for all applications. This is why undertaking appropriate testing/screening of products is an important precursor to making a final choice of product to use.

Effective inhibition is most readily achieved and maintained in a system that is kept clean, e.g. not heavily scaled or fouled; with an absence of waxy deposits/films and no build-up of standing solids. Cleaning is most efficient when carried out mechanically but may also benefit from the aid of chemical surfactants. Depending on operating system history, cleaning may be an initial and/or periodic supporting requirement.

8.1.1 Key Benefits

The benefits of chemical corrosion inhibition are numerous and can be summarised as:

- Cost-effectiveness – this can be sensitive to how the economics are run but generally it is the most capital expenditure (CAPEX) efficient option for corrosion mitigation.
- Extends use of materials with established and favourable engineering properties – primarily talking about use of CLASs.
- Flexible response – it can be adjusted to meet changing operating conditions.
- Retrofit treatment – it can respond to unexpected increase in system corrosivity, a desired change in service conditions and/or life extension.
- Assurance – allowing confidence where there is limited access and opportunity for inspection and/or corrosion monitoring.
- Can inhibit metal loss corrosion due to sweet and sour conditions.

A detailed treatise on oilfield corrosion inhibitors can be found elsewhere [1].

8.1.2 Inhibitor Formulation

Formulation of a product is manipulated in order to provide a range of performance characteristics depending on the nature of the application and service conditions. These include:

- ability to reliably maintain an acceptable inhibited corrosion rate;
- strong surface adsorption (often synergistic between the components); long-lasting film for batch applications;
- solubility/dispersibility in the different phases present for effective deployment/ transport throughout a system
- low toxicity to humans and environmentally friendly (e.g. following OSlo and PARis, OSPAR, regulations in Europe);
- high flash point for reduced hazard when handling;
- low viscosity for transport in umbilicals;
- temperature stable for storage in hot (60 °C) and cold (-40 °C) climates;
- compatible with elastomers that are used in the system, i.e. do not cause excessive swelling, cracking, or hardness changes;
- non-corrosive to metals on storage and application of the undiluted material – may require the use of stainless steel lines and internally lined storage vessels;

- compatible with other chemicals used in the system, e.g. hydrate inhibitors, scale inhibitors, scavengers, and biocides;
- compatible with chemicals with which they are to be mixed, e.g. surfactant packages;
- non-foaming;
- non-emulsifying;
- persistency, both in terms of chemical stability, due to other competing chemicals, as well as physical from, for example, surface shear due to liquid flow and turbulence.

It may not be possible (or necessary) to satisfy all of these parameters in one product, in which case a compromise needs to be produced that considers the most important characteristics or fits all the characteristics for a limited application. Some deficiencies may be overcome by alternative procedures, such as the use of trace heating lines in cold climates.

Figure 8.1 shows the typical component parts of a fully formulated corrosion inhibitor. Chemical service companies may well have proprietary or patented technologies that they incorporate into their formulations. Surfactants, co-solvents, and neutralisers are also added to aid in dispersion and transport.

Formulations can also contain an inorganic or organic derivative sulphur species at low concentration. It is unclear how these additives function: they may decompose to form a thin protective iron sulphide layer or strongly adsorb at the metal interface, forming bonds similar to chelated structures. However, this practice is far less common today as there are doubts about the actual benefits when used in the field versus a possible benefit of merely enhancing product performance under laboratory test

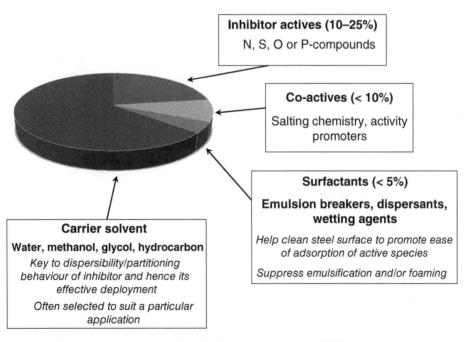

Figure 8.1 Typical component parts of a fully formulated corrosion inhibitor.

conditions. Also, if in fact it does promote formation of an iron sulphide film inhibitor, under-dosing or poor deployment may promote pitting.

It can be seen from Figure 8.1 that the actual percentage of active inhibitor species present in a formulated product may be no more than 25%. Therefore, for example, for a specified optimal formulated product concentration of 50 ppm (based on total produced fluids), the resulting concentration of active inhibitor species present in the fluids may range from 5 to 12 ppm.

8.1.3 Inhibitor Species and Functionality

The principal role of a corrosion inhibitor molecule is to inhibit the reaction between a steel surface and its environment in a definable, reproducible, and controllable manner. Inhibitors used to treat produced fluids and gas typically have the following attributes:

- They reduce the uninhibited rate of corrosion to an acceptable and continuously manageable level through formation of an adsorbed surface film.
- Corrosion is not totally suppressed – this is a consequence of how tightly the adsorbed inhibitor molecules can sterically pack on a surface. Inhibitor efficiencies (described in Section 8.1.4) typically fall in the range 95–99% for well-managed, well-maintained, and clean systems and therefore an inhibitor is commonly used together with a corrosion allowance[1] designed into the wall thickness, certainly for pipelines.
- For application by continuous injection into the production stream, inhibitor film persistency is dependent on the inhibitor species always being present, preferably at an optimal concentration in the corrosive phase to give maximum efficiency (lowest inhibited corrosion rate) without introducing a change in corrosion morphology (not promoting pitting).

The vast majority of oilfield active inhibitor species used today are organic-based fatty acids and amines, examples of which are shown in Figure 8.2. They are almost

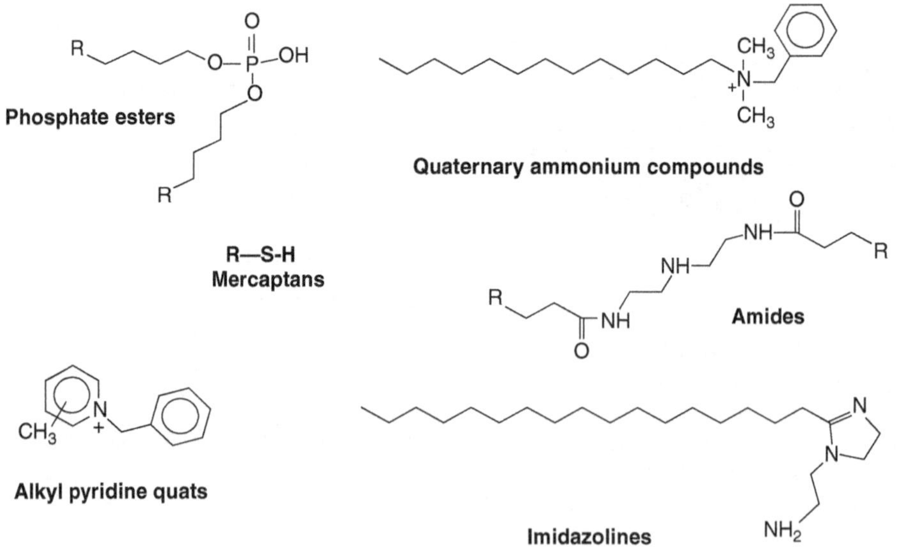

Figure 8.2 Typical corrosion inhibitor active species.

exclusively developed to inhibit the corrosion of CLASs – the main exception being acidising of wells when corrosion-resistant alloys (CRAs) are deployed.

8.1.3.1 Functionality

Corrosion inhibitors are principally used for treating produced fluids and gas and their performance characteristics can be summarised as follows:

- They work under anaerobic conditions acidified by the presence of CO_2 and/or H_2S.
- Even the presence of low parts per billion (ppb) levels of O_2 can have an adverse effect on performance and promote pitting [2–4]: the exceptions are quaternary ammonium compounds, that can also be applied to water injection systems due to a tolerance for the presence of dissolved oxygen.
- They typically work in the pH range 3.0–6.5; outside this range, performance and choice may be limited.
- Inhibition is achieved by adsorption to the steel surface via physisorption (charge attraction between inhibitor molecule and steel surface) and chemisorption (semi-chemical bonding) [5]
- Performance is significantly challenged at temperatures >150 °C.

As mentioned earlier, the resulting adsorbed inhibitor film is a mono-molecular layer thick. However, inhibitor film continuity over a surface and performance (cf. inhibitor efficiency) can be affected by surface cleanliness and roughness.

It should also be noted that the inherent surface active nature of inhibitor species means that they may equally adsorb on to other (more) favourable surfaces present – e.g. corrosion product ($FeCO_3$ and FeS) films or sand particles. The significance of adsorption on to corrosion product films is far from clear but there appears to be no evidence of it having a detrimental effect per se on inhibitor performance. However, sand particles, which generally have a very high surface area-to-volume ratio, can significantly reduce the active concentration of inhibitor present, depending on the nature and level of sand present and whether it is standing or mobile. Furthermore, if standing, this will shield or impair inhibitor access to the underlying steel surface; whereas, if slow moving along the bottom of line, for example, sand may continuously disrupt inhibitor film stability, resulting in loss of inhibitor efficiency.

8.1.4 Inhibitor Performance

Inhibitor performance is commonly expressed and measured in terms of efficiency – i.e. the percentage amount an inhibitor is able to reduce the uninhibited corrosion rate when present at a given concentration in solution. This is simply written as Eq. (8.1):

$$\%\text{inhib}_{\textit{eff}} = \left[\frac{CR_{\text{uninhib}} - CR_{\text{inhib}}}{CR_{\text{uninhib}}} \right] \times 100 \tag{8.1}$$

where:

$\%\text{Inhib}_{\text{eff}}$ = inhibitor efficiency (%)
CR_{inhib} = inhibited corrosion rate
CR_{uninhib} = uninhibited corrosion rate

Given the organic structure and molecular size of an inhibitor molecule, it is highly unlikely to exceed an inhibitor efficiency >99% due to a spatial limitation of how well they can closely pack together on a steel surface in forming a monolayer: i.e. they cannot achieve 100% surface coverage. Also it is a common expectation at the design stage that an inhibited corrosion rate of 0.1 to ≤0.5 mmy^{-1} is consistently achievable. This 'residual' inhibited corrosion rate then has to be safely accommodated when considering tubing and pipe wall thickness versus design life or an acceptable replacement life. For pipes, this is addressed through specifying a corrosion allowance (CA) additional to the required pipe wall thickness governed by the mechanical design code [4] for maximum operating pressure containment, strength, defects, etc. Factors such as pipe diameter, inspection sensitivity, welding, steel product supply, and availability and cost will limit the actual CA possible. The picture downhole is generally governed by required strength and weight of tubing versus design and complexity of a well completion and deliberately introducing a CA is generally not a practical or desirable option.

As the long chain hydrocarbon tail (cf. Figure 8.2) of an inhibitor molecule is hydrophobic and likely oleophilic, it may under favourable conditions establish a hydrocarbon film interlaced with the adsorbed inhibitor film and hence afford enhanced inhibitor efficiency. Reliance on this for design purposes is, however, questionable, whereas it may present an additional benefit in practice.

Inhibitor efficiency is a simple and convenient engineering parameter to compute, but it is important to recognise that its value is highly sensitive also to how the uninhibited corrosion rate is determined. The latter can be determined by predictive modelling (at the design stage), measurement in the laboratory (in support of commissioning and through life inhibitor selection and QA/QC) and, maybe measurement in the field when on stream. However, if a field is designed to operate with an inhibitor from day one, then field measurement will likely be impractical to undertake early in the life of a field when water production may be very low.

Use of inhibitor efficiency once a field is in operation has per se limited value. Of greater practical importance is monitored inhibited corrosion rate and maintaining the prescribed inhibitor injection rate to ensure optimal concentration in the produced fluids – the concept of inhibitor availability discussed later under inhibitor application.

8.1.4.1 The Effect of Fluid Flow

Liquid velocity is seen as an important factor affecting inhibitor performance. Commonly there is an upper liquid velocity limit placed on treatment with a corrosion inhibitor (e.g. 20 m s^{-1}). However, it is increasingly apparent that the picture is less clear-cut than has been presented in the past. The practicality of being able to use more sophisticated in-situ and ex-situ surface analysis techniques has revealed that in fact a well-established adsorbed inhibitor film can be very resilient to disruption by flow alone.

The appropriate flow parameter that can be translated directly between laboratory and service-generated surface conditions is liquid shear stress. The aforementioned 20 m s^{-1} flow velocity for water through a 10 in. diameter pipeline equates to a surfcae liquid shear stress of ~325 Pa, whereas the performance of inhibitor films have been shown to be unaffected when exposed to shear stresses >1000 Pa – admittedly this is starting with a clean (polished) steel surface under laboratory conditions. Nevertheless, it would be folly not to consider the effect of flow on inhibitor performance as part of a precursor screening and selection test programme before going to field deployment. This should cover shear stresses representative of both bulk and localised operating system flow conditions.

8.1.5 Environmental Acceptance

The inhibitor chemistries alone, shown in Figure 8.2, before considering final product formulation, can be a challenge to meeting all environmental requirements and legislation. The push to identify and develop 'greener' inhibitor molecule chemistries remains slow outside of modelling and the laboratory [7, 8]. Commercial and end user pull for 'greener' chemistries is affected by a number of factors. Until there is much broader take up, 'greener' chemistries generally carry a higher unit (premium) price than conventional inhibitor chemistries. Also there is question mark against their ability to deliver corrosion inhibitor performance (efficiency) matching or bettering that of the well-established chemistries – a performance benchmark the industry has come to depend on. And balancing biodegradability versus maintaining useful in-service inhibitor life activity [9] remains a challenge.

Environmental legislation is an important factor in the final selection of a formulated corrosion inhibitor product – as too it affects the acceptable use of all production chemicals. The North Sea is generally seen as having the most complex requirements, as detailed within the OSPAR (unified OSlo and PARis Conventions) guidelines. Assessment under these guidelines results in a chemical being ranked according to its suitability to be deployed offshore. The field operator must then seek a permit from the authorities for the controlled discharge of the chemical within limits over a time-limited period. Three eco-toxicological tests are required by OSPAR: acute toxicity; bioaccumulation; biodegradation in sea water.

In the United States, environmental regulations are administered by the Environmental Protection Agency (EPA). They differ from the North Sea and also vary for different locations within the US jurisdiction. A notable difference in approach is the application of a critical dilution factor (CDF) assigned to produced water discharges for each facility in assessing the toxicity of each production chemical to marine organisms.

For regions where environmental regulations are less well defined, it is often common for local regulatory agencies or oil companies to request chemical service companies to provide products that comply with North Sea requirements.

Environmental issues, considerations, and regulations, as well as greener chemicals, are discussed in detail for all production chemicals by Kelland [10].

8.2 Inhibitor Testing and Application

The complexity of the service conditions, the criticality of overall inhibitor product formulation to achieving optimal performance for the application, and the heavy reliance of the design case or basis of design (BoD) on effective inhibitor treatment through its life in the field mean that undertaking an appropriate selection programme is an important step. The programme will have elements of product screening to then focus in on testing under critical application-specific conditions prior to limited or field-wide deployment.

Formulated inhibitor products are commonly grouped under one of the following descriptions:

- oil-soluble – continuous/batch injection;
- water-soluble – continuous injection;
- oil-dispersible (controlled solubility) – continuous or batch injection;
- combination of the above (e.g. oil-soluble/water-dispersible or vice versa).

However, strict adherence to a definition of being truly soluble versus dispersible can be a grey area. Care needs to be exercised as, irrespective of classification, a formulated inhibitor will separate between oil and water: the oil/water ratio will influence the actual concentration of inhibitor in the aqueous phase even if a formulated product is designated as water-soluble. This partitioning behaviour needs to be factored in when determining the field injection rate with a product's partition coefficient determined, using a field sample of additive-free crude oil. Furthermore, for certain applications – e.g. downhole and long-distance subsea completions – the use of combination products (e.g. corrosion/scale inhibitors) may be chosen for ease and economy of design and operation of injection facilities.

8.2.1 Operating Conditions

The operating conditions and constraints under which the inhibitor is required to work must be understood and defined, not least because:

- It will determine the methodology and scope of the test programme required for sound inhibitor selection.
- If the operating system to be treated has already experienced a level of corrosion damage, this can be more challenging to manage to an acceptable level of inhibition:
  - Treating localised corrosion may be more demanding – i.e. require a higher chemical dosage than under less energetic surface flow conditions, at least for an initial period.
- The flow regime will influence the effectiveness of deployment in the field and influence the location of potential corrosion hot spots, such as sharp bends or dead legs.
- System cleanliness can be a significant factor on inhibitor performance.
- It will determine the criticality and order of importance of factors affecting performance, for example:
  - If compatibility with another production chemical is likely to be an issue, then for which one can a compromise on performance be accepted if one is unable to find a fully compatible optimum solution?
  - Legislation or operator policy requires the use of inhibitors with a particular environmental rating.

8.2.2 Inhibitor Testing/Selection

In recent years, much effort has been directed at the development of test methods for studying and evaluating corrosion inhibitor performance. This has resulted in increased sophistication, especially when combined with use of ex-situ and in-situ surface analysis techniques, while being manageable and not excessively expensive to establish and run in a conventional laboratory environment. The most commonly used laboratory test methods are summarised in Table 8.1 and further described elsewhere [1].

The detail and practice and pros and cons of inhibitor testing constitute a book in their own right. Those listed in Table 8.1 can be viewed from top to bottom as increasing in practicality and sophistication, depending on the primary field conditions that need to be reproduced in the laboratory. Which of the listed methods used will depend on the primary field conditions that need to be reproduced in the laboratory. Tests are typically run over a 1–7-day period and ideally in triplicate. All tests are conducted in 1–5 l glass vessels – perhaps up to 10 l for a glass flow loop – with the stirred rotating cylinder electrode (RCE) and rotating cage test cell arrangements lending themselves most readily to elevated temperature and pressure build and operation (important if the in-field partial pressure of an acid gas exceeds 1 bara and for temperatures >100 °C). The detail behind running the above test methods is addressed by ASTM [11], NACE [12] and UK HSE [13] and usefully reviewed by Papavinasam and Revie [14].

8.2.2.1 The Media
With the exception of the wheel test and the rotating cage, all the test methods are normally operated with only an aqueous phase present: an essential requirement when making electrochemical corrosion rate measurements. However, the aqueous phase can be preconditioned separately in contact with a hydrocarbon phase before being introduced into the test cell. It is possible to precondition the test electrode/coupon arrangement with a mixed hydrocarbon/aqueous phase, although this requires a more involved, but manageable, procedure. In both cases it may well not be possible to source an additive chemicals-free sample of field crude oil, so it may be necessary to resort to using suitable inert synthetic oil to represent the hydrocarbon phase. The representativeness of the latter needs careful review and preferably is following the advice of an expert production chemist and flow assurance engineer.

8.2.2.2 Appropriate Tests
Chemical service companies will run all or a number of test methods outlined earlier (see Table 8.1) in support of inhibitor development and selection for customer-specific applications. These test methods can also readily be accessed through independent technical service companies and at certain universities. All are suited to conducting inhibitor screening across the product range of one or several service companies, but the latter will usually require engaging an independent technical service company because of individual commercial product confidentiality. Whether one or several of the test methods match generating data suitable for making a final selection for field deployment will depend on the complexity of the field conditions.

If field conditions are particularly complex and the facilities are of high criticality with respect to service, safety or environmental threat, then it may be deemed necessary to conduct final inhibitor testing and selection using a large diameter flow loop that is able to simulate multiphase flow. This is a costly, time- and labour-intensive and a specialised step with only a small number of such facilities available worldwide, notably at: the Institute for Corrosion and Multiphase Technology, Ohio University, USA; the Erosion/Corrosion Research Center, the University of Tulsa, USA; and the Institutt for Energiteknikk (IFE), Norway.

Table 8.1 The most commonly used laboratory test methods for corrosion inhibitors.

Corrosion inhibitor test method	Description	Liquid shear stress pascal (Pa)	Data	Advantages	Limitations
Wheel/bottle test	Coupons inside sealed bottles containing conditioned inhibited fluid mounted on spokes of a wheel	Limited agitation generated by rocking action by rotating wheel	Corrosion rate (CR) from weight loss measurement	Ease of setting up and use with low cost and portable for in-field use	Very basic test that poorly represents actual conditions. Now rarely used
Stirred test cell (bubble test)	Test electrode coupon mounted in sealed vessel and exposed to stirred pre-conditioned inhibited aqueous phase	~ 1 Pa (pipeline fluids velocity typically <1 m s^{-1})	CR from LPR or other electrochemical methods. Also suitable for weight loss measurement.	Ease of set-up, use and low cost. Flexible to considering influence on inhibitor performance of presence of localised forms of corrosion – e.g. pits, preferential weld corrosion (PWC), under deposite corrosion (UDC)	Poor control and definition of surface hydrodynamic conditions that can affect reproducibility
Rotating cylinder electrode (RCE)	Vertically orientated rotating cylindrical test electrode mounted in sealed vessel and exposed to pre-conditioned inhibited aqueous phase	Up to ~80 Pa (pipeline fluids velocity approaching ~10 m s^{-1})	CR from LPR or by other electrochemical methods	Ease of set-up and relatively low cost. Most commonly used for lab screening and often field selection of inhibitors under well-defined and reproducible surface hydrodynamic conditions	Not very flexible to considering influence on inhibitor performance of presence of localised forms of corrosion – e.g. pits, PWC, UDC
Rotating disc electrode (RDE)	Horizontally orientated rotating disc test electrode mounted in sealed vessel and exposed to pre-conditioned inhibited aqueous phase	Varying shear stress across the disc surface depending on size/configuration		Steady state surface conditions quickly attained with high reproducibility, especially where diffusion is important factor	More suited to fundamental electrochemical studies. Rarely used for routine inhibitor testing

Small diameter flow loop	Typically up to 30 mm diameter sealed pipe loop with flush mounted test electrodes / coupons exposed to recirculating conditioned inhibited test fluid	Up to ~250 Pa shear stress (pipeline fluids velocity approaching ~15 m s^{-1})	CR usually from LPR or by other electrochemical methods; but suitable for weight loss measurement too	Most suitable for pre-conditioning of test sample with mixed inhibited hydrocarbon / water phase and testing threat of PWC	Capital cost, size and operating complexity
Jet impingement test cell	Often combined with RCE or possibly RDE using recirculating jetted pre-conditioned inhibited aqueous phase on to test electrode	Up to ~350 Pa shear stress (pipeline fluids velocity > 20 m s^{-1} and where subject to turbulent conditions)	CR usually from LPR or by other electrochemical methods	Can generate well defined very aggressive test conditions	Added complexity of design and specialist operation. Not commonly used
Rotating cage test cell	Coupons mounted vertically to form a rotating cage configuration exposed to conditioned inhibited test fluid in sealed vessel	Not easy to determine shear stresses generated but can simulate erosive conditions caused by droplet impact	CR from weight loss measurement	Ease of design and set-up and able to generate aggressive test conditions	No definition of surface conditions that can affect reproducibility

8.2.2.3 Complementary Considerations

In the presence of sand, with the likelihood of forming standing accumulations and/or erosion-corrosion occurring, the role and efficacy of oilfield corrosion inhibitors become less clear. In the case of the latter, it raises the question: can an inhibitor have any material effect on the erosion process and exhibit the same efficiency on the corrosion process as when erosion is absent? This continues to be researched in some detail at the University of Leeds [15] and reference to this work is recommended. Whereas, if there is a risk of standing accumulations of sand occurring, use of the novel test cell arrangement developed by the UK National Physical Laboratory (NPL) [16, 17] is strongly recommended.

One further important consideration affecting the selection of a corrosion inhibitor is the potential for preferential weld corrosion (PWC) to occur: essentially a galvanic mechanism between the weld, the heat-affected zone (HAZ), and the parent material that can result in severe pitting to the weld or HAZ. PWC has been experienced in the presence of produced fluids where here it is best controlled by treatment with a suitable corrosion inhibitor. (By contrast, using a high Ni welding consumable has proved successful for mitigating PWC in water injection systems, where dissolved oxygen is the primary driver of the galvanic action.) It is, therefore, necessary to assess if there is a credible threat of PWC and, if so, include it in the test methodology for final inhibitor selection [18]. It should also be noted that if due attention is not paid to this requirement, it may exacerbate the risk through selection of an inappropriate inhibitor.

Moving to the next step of conducting a 'field trial' may well be limited in its entirety and subject to the capability that already exists. It may only be feasible to deploy inhibitor selectively to one part of a field as a trial. It may be possible to introduce more sampling points and temporary installation of a side-stream incorporating additional corrosion monitoring instrumentation, and introduce a suitably revised and targeted inspection programme. However, all this will depend on the infrastructure and the operating complexity of the field and critically the approval and support of operations. In reality, it is likely that a 'field trial' would need to run for a minimum of six months and more likely a year; and may well not be a feasible consideration at all at the start-up and during early life of the field when water cut is very low. Consequently, moving directly into full system deployment of an inhibitor with appropriate performance monitoring is the most common step after laboratory selection together with ongoing 'live' performance optimisation and annual performance review.

8.3 Inhibitor Application/Deployment

Having expended time and effort on identifying the technically best or most cost-effective inhibitor to treat a system, if it is not applied consistently and conscientiously, the expected performance will not be realised, leaving the system running with a higher corrosion risk than designed, expected, or desired. Application is not just about the method but how it is then managed: having chosen inhibition as the primary corrosion control measure, it is a 'cradle-to-grave' commitment and certainly not a 'fit-and-forget' option. This is further discussed in Chapter 18.

8.3.1 Continuous Injection

As already mentioned, this is by far the most common and generally the most cost-effective approach. It is a decision normally taken at the design stage and requires installation, commissioning, and ongoing management and maintenance of inhibitor injection facilities together with well-developed operating procedures, key performance indicators (KPIs), and sound inhibitor supply logistics. It ideally requires early engagement with chemical service companies to optimise injection system design and subsequent operation versus inhibitor selection.

8.3.1.1 Inhibitor Availability

As touched on earlier, the concept of inhibitor availability [19] has an important role to play both at design stage and during application. Inhibitor availability (A) is the percentage time the inhibitor is present at a concentration to give or improve a defined (or desired) inhibited corrosion rate. For design purposes, it is common to assume an inhibited corrosion rate (CR_{inhib}) of 0.1 mmy^{-1}; or it can be computed from the uninhibited corrosion rate ($CR_{uninhib}$) multiplied by the determined inhibitor efficiency (see Eq. 8.1). Therefore, the actual inhibited corrosion rate (CR_{actual}) achieved when treating with a corrosion inhibitor is shown in Eq. 8.2:

$$CR_{actual} = CR_{inhib} \times \left\{ \frac{A}{100} \right\} + CR_{?inhib} \times \left\{ \frac{100 - A}{100} \right\} \tag{8.2}$$

The value of $CR_{?inhib}$ will depend on inhibitor film persistency should the injection go down or if under-dosing occur for whatever reason, where the worst (default) case would be $CR_{?inhib} = CR_{uninh}$. Clearly the aim is to always achieve 100% injection at the required concentration, but this may not be consistently achieved for various operational reasons, e.g. faulty injection pump or partial blockage of inject line; logistical upset in fresh supply of inhibitor; lagging response to a change in operating conditions, such as water content of produced fluids; a lag in monitored corrosion rate, especially if only using weight loss coupons.

Equation 8.2 is also important in determining the inclusion of a suitable CA for a given design life. Figure 8.3 shows a design case example of the relationship between

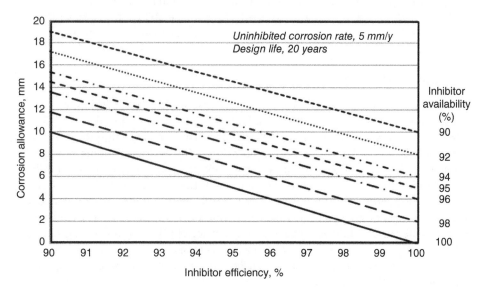

Figure 8.3 A design case example of the relationship between inhibitor efficiency and availability versus corrosion allowance.

inhibitor efficiency and availability versus CA, assuming an inhibited corrosion rate of $0.1\,mmy^{-1}$ and worst (default) case of $CR_{?inhib} = CR_{uninhib}$. An inhibitor availability of 98% equates to an inhibitor injection downtime or under dosing of 7 days in a year and, 95% equates to 18 days in a year.

Once operational, there is the opportunity to measure CR_{actual} through corrosion monitoring and inspection, accepting that it is a lagging measurement, typically expressed as an annualised value. From this, a revised value for A consistent with the remaining CA can be computed to give maximum inhibitor downtime and CR_{actual} targets (KPIs) set for the next year, to be reviewed annually. This can also become an increasingly important consideration if the need or desirability to continue operating beyond the original design life arises, which is not an uncommon factor arising in later life of a field.

8.3.2 Field Evaluation

While corrosion monitoring and inspection are key activities in assuring the required level of inhibitor performance is being achieved, complementary indirect measurements also have an important role to play, especially given the former are lagging indicators. The latter include: conducting periodic analysis of water samples, ideally sampled across a system, for concentration of active inhibitor species, in-situ pH, level and type of any solids present; presence of bacterial activity; continuous monitoring of set inhibitor injection rate versus required system demand, through each injection point, based on total fluids and factoring in the effect of changing water cut; total inhibitor volume usage versus whole system demand. Day-to-day operating performance indicators (PIs) and need for additional KPIs can be set from the capture of complementary indirect data. These can then be viewed on a weekly/monthly dashboard display – via a data management system – for review and management purposes. These measures are touched upon again in Chapters 10 and 18.

It should be noted that if considering use of a co-formulated product – e.g. a combined scale and corrosion inhibitor – this will affect the flexibility of adjusting the injection rate to subsequently meet the changing requirements of either inhibitor.

8.3.3 Wet Gas Lines

Wet gas lines can be treated with continuous injection of corrosion inhibitor although the usually low or limited volatility of inhibitors can present a problem of effective deployment throughout a system if physically unaided – namely, inhibitor drop out to bottom of line a limited distance from an injection point. Periodic running of a suitably designed mechanical pig through a line may well be required to carry the inhibitor forward and disperse it 360° around the pipe wall – the latter a particular requirement if there is a susceptibility to top-of-line corrosion. When a gas line is operated in dense phase conditions, finding an inhibitor formulation that is physically stable under such conditions may be a significant challenge – risk of solvent flashing off, leaving behind a sticky black deposit that can, for example, clog up screens. The better option is to ensure the water content of the gas is meticulously controlled so that the line temperature never falls below the water dew point.

8.3.4 Downhole Inhibition

Continuous injection of inhibitor downhole is generally most effectively achieved by inclusion of a capillary injection string as part of the well completion design. It may also be possible to undertake via a gas lift system if present or directly via the annulus if suitable well design and downhole valving is available. However, both are less easy to control in achieving a prescribed injection rate and can be prone to gunking up of injection valves.

8.3.4.1 Batch Treatment

This method of application is now almost exclusively limited to downhole treatment of wells. There are three main methods, all of which require the well to be shut in while the inhibitor application is undertaken:

- Most common: tubing displacement of reservoir fluids with inhibitor batch to bottom of well – leave to soak and coat internal surface of wetted tubing and rely on film persistency under producing conditions until next treatment.
- Less common: recirculation between well bore and annulus with latter filled with inhibitor – inhibition works in a similar way to tubing displacement.
- Rare: squeeze inhibitor into producing zone where it loosely adsorbs and flows back over time with the well back on production – could be viewed as a quasi-continuous treatment except inhibitor flow back is not uniform, the bulk of inhibitor flowing back within hours of bringing a well back on production.

The first two methods rely exclusively on inhibitor film persistency where product formulation looks to use not only adsorption of inhibitor but also limiting inhibitor film solubility in the produced fluids. This type of application is seen as producing a thicker, albeit less uniformly structured, film than due purely to adsorption – cf. adsorbed monomolecular layer formed under continuous injection of inhibitor – but places a high reliance on knowing how long acceptable film persistency will be maintained before retreatment is required: this may be a matter of a one or two weeks to a month depending on the severity of the well conditions. The logistics of conducting a treatment versus required frequency can be a challenge, depending on location, the size of the field, and the ease of individual well access. Working closely with a chemical service company offering this type of treatment is an essential element in its success and in determining the actual application detail for any given well configuration. For fields in remote and environmentally sensitive locations, it is increasingly common to go the CRA route downhole if corrosion presents a credible threat.

8.4 Summary

The use of corrosion inhibitors to mitigate the threat of internal corrosion affecting the design and operation of upstream oil and gas production infrastructure and facilities has proved to be a successful, resilient, flexible, and cost-effective means of managing that threat over the past 60+ years. Arguably it has become *the* base case consideration enabling and extending the use of CLASs. However, with new projects exploiting reservoirs with fluids of increasing complexity and more aggressive environments, consistently achieving high inhibitor efficiency of performance and application will be paramount to economic project delivery and integrity risk management [20].

Commercial corrosion inhibitors are formulated products, where formulation is a critical element affecting the efficacy of the performance versus specific application. This is why there is not, at least so far, a 'silver bullet' corrosion inhibitor suitable for all applications; and why undertaking appropriate testing and screening of products is an important precursor to making a final choice in achieving successful field application.

The history of using corrosion inhibitors has not been without its troubles, certainly in the early days when a sound fundamental understanding of how they function lagged behind demand. However, issues, should they arise now, generally are associated with insufficient or poor attention to managing inhibitor application and not inhibitor performance per se in the ability to reduce corrosion to an acceptable level if applied properly.

Corrosion inhibitor application is not just about the method but how it is then managed: it is a 'cradle-to-grave' commitment and certainly not a 'fit-and-forget' option.

Note

1 It should be noted that depending on corrosion philosophy, CA may be considered a safety precaution for upset conditions rather than additional wall thickness to be consumed over the years.

References

1 Palmer, J.W., Hedges, W., and Dawson, J.L. (eds.) (2004). *The Use of Corrosion Inhibitors in Oil and Gas Production*, EFC Publication No. 39. London: W S Maney and Son Ltd.

2 Edwards, A., Osborne, C., Webster, S. et al. (1994). Mechanistic studies of the corrosion inhibitor oleic imidazaoline. *Corrosion Science* 36 (2): 315–325.

3 Jovancicevic, V., Ramachandran, S., and Prince, P. (1999). Inhibition of CO_2 corrosion of mild steel by imidazolines and their derivatives. *Corrosion* 55 (5): 449–455.

4 Gulbrandsen, E., Kvarekvål, J., and Miland, H. (2005). Effect of oxygen contamination on inhibition studies in carbon dioxide corrosion. *Corrosion* 61 (11): 1086–1097.

5 Durnie, W.H., Kinsella, B.J., de Marco, R., and Jefferson, A. (2001). A study of the adsorption properties of commercial carbon dioxide corrosion inhibitor formulations. *Journal of Applied Electrochemistry* 31 (11): 1221–1226.

6 ASME, Pipeline transportation systems for liquid hydrocarbons and other liquids. ASME B31.4, 2006.

7 I R Collins, B Hedges, L M Harris, et al., The development of a novel environmentally friendly dual function corrosion and scale inhibitor. Paper presented at SPE International Symposium on Oilfield Chemistry, 13–16 February, Houston, Texas, 2001.

8 A Jenkins, A Fraser, L Nolasco, et al., Development of environmentally acceptable combined scale/corrosion inhibitor for a North Sea oilfield with high velocity flow. NACE Annual Corrosion Conference, Paper No. 01257, 2012.

9 D G Hill, Testing of green inhibitors for corrosion inhibition and toxicity. NACE Annual Corrosion Conference, Paper No. 04409, 2004.

10 Kelland, M.A. (2014). *Production Chemicals for the Oil and Gas Industry*, 2e. CRC Press.

11 ASTM, Standard guide for evaluating and qualifying oilfield and refinery corrosion inhibitors in the laboratory. ASTM G 170-06, 2012.

12 NACE, Laboratory test methods for evaluating oilfield corrosion inhibitors: technical committee report, Task Group T-1D-34, Item No. 24192. NACE International Publication 1D196, 1996.

13 J Hobbs, Reliable corrosion inhibition in the oil and gas industry, RR1023 Research Report, UK Health and Safety Executive, 2014.

14 S Papavinasam and RW Revie, Testing methods and standards for oilfield corrosion inhibitors, NACE Annual Corrosion Conference, Paper No. 04424, 2004.

15 Neville, A. and Wang, C. (2009). Erosion–corrosion mitigation by corrosion inhibitors: an assessment of mechanisms. *Wear* 267: 195–203.

16 Hinds, G. and Turnbull, A. (2010). Novel multi-electrode test method for evaluating inhibition of underdeposit corrosion. Part 1: sweet conditions. *Corrosion* 66 (4): 046001–046001-10.

17 Hinds, G. and Turnbull, A. (2010). Novel multi-electrode test method for evaluating inhibition of underdeposit corrosion. Part 2: sour conditions. *Corrosion,* 66 (5): 056002–056002-6.

18 R Barker, X Hu, A Neville, and S Cushnaghan, Assessment of preferential weld corrosion of carbon steel pipework in CO_2-saturated-flow-induced corrosion environments, NACE Annual Corrosion Conference, Paper No. 01286, 2012.

19 B Hedges, D Paisley and R Woollam, The corrosion inhibitor availability model, NACE Annual Corrosion Conference, Paper No. 00034, 2000.

20 A Crossland, R Woollam, J Vera, et al., Corrosion inhibitor efficiency limits and key factors, NACE Annual Corrosion Conference, Paper No. 11062, 2011.

9

Coating Systems

Coating and painting systems are the most common forms of external corrosion control used in the oil and gas industry, offering a cost-effective and feasible barrier to mitigate corrosion threats. What differentiates the terms 'coating' and 'painting' is both in relation to the thickness and purpose of the system deployed. Painting applications are generally very thin, no more than a few hundred micro-metres thick at most, limited to corrosion mitigation in atmospheric conditions and usually serve a decorative, as well as protective, use. Coating applications are generally much thicker (can be several millimetres thick) and used both for atmospheric, subsea, and underground protection. A third term, 'lining', is generally referred to thick coatings, or solid non-metallic materials, used for internal protection of storage tanks, separators, downhole tubular, and pipelines intended to afford a more durable corrosion barrier. Coatings applied internally for corrosion mitigation also offer improving flow dynamics, reducing inhibitor cost, and externally to offer improved appearance, minimising corrosion, and commonly to enhance implementation of cathodic protection.

There are numerous coating and painting standards, produced by international organisations and companies, covering every type of coating for different types of environment. These standards cover several aspects of coating selection from environment suitability to qualification tests, application, transport, and storage of coated pipes and inspection. However, even with so many standards available, the process of coating selection, testing, application and inspection is far from a routine assignment and requires considerable attention to detail.

This chapter summarises current industry practice in the use of non-metallic coating systems and outlines shortfalls, advantages, and methods used in mitigating corrosion threats in hydrocarbon production activities. The content excludes metallic lining and cladding which is dealt with in Chapters 2 and 15.

9.1 External Pipeline Coatings

The application to pipelines is arguably the most critical and important use of coatings. Such coatings can cover kilometres of pipelines, making its application, testing, inspection, quality assurance and control (QA and QC) somewhat challenging. A small pinhole or 'holiday' – as it is referred to in coating terminology – in a long pipeline can

Corrosion and Materials in Hydrocarbon Production: A Compendium of Operational and Engineering Aspects, First Edition. Bijan Kermani and Don Harrop.
© 2019 John Wiley & Sons Ltd. This Work is a co-publication between John Wiley & Sons Ltd and ASME Press.

lead to a leak and uncontrolled release of oil and gas. This can cause shutdown and lost production, not to mention environmental and health and safety concerns.

Although international coating standards generally cover a specific type of coating, these standards are not usually environment-specific. For this reason, it is best for each oil and gas company to have their own internal standard, taking in the relevant information from international standards, while making it specific to their operating environment.

Here, only epoxy and polyolefin type coatings are considered as they are the principal types of coating presently used widely in pipeline applications. While used in the past, coal tar enamel has now been banned in many countries due to health, safety, and environment (HS&E) concerns.

9.1.1 Fusion-Bonded Epoxy (FBE) Coating

Fusion-bonded epoxy (FBE) is by far the most commonly used pipeline coating. FBE coatings are available for a wide range of temperatures, have good adhesion to steel, and require relatively thin coats. They have good chemical resistance, reasonable resistance to mechanical damage, and are easy to repair. In underground pipelines, where FBE coating on pipelines is supplemented by cathodic protection (CP), in the event of any coating damage, in most cases the CP current is able to travel to the bared location and prevent corrosion.

FBE is applied by an electrostatic spray method. It can also be applied in the field for field/weld joints. FBE can also be used for internal coating of pipelines and vessels. There are special high temperature versions of FBE that can be used at temperatures of 150 °C or even higher. However, the high temperature FBEs are less flexible than the standard FBEs and also more costly.

In areas where the pipeline coating is prone to be impacted by constituents in its environment, such as pipelines laid in the sand, two or more layers of FBE can be used. The properties of the different FBE layers can be designed differently to allow more resistance against impact with durability in mind. For example, the first/inner layer can be a standard FBE employed for its corrosion protection properties while the second/outer layer could have more erosion-resistant properties.

A common cause of failure of FBE coatings is uneven heating of the pipeline during coating application. This requires careful QA/QC.

9.1.2 Polyolefin Coatings

Polyolefin coatings, mainly polyethylene (PE) and polypropylene (PP), have gained growing popularity through the years in their use as pipeline coatings. These are normally used as three-layer coating systems. The first layer is FBE, which acts as the primary corrosion protection coating, bonded directly to the pipe wall. The second layer is an adhesive, tying the FBE to the outer polyolefin coating. The latter is a relatively thick layer of polyolefin (normally > 1 mm), offering impact resistance and some insulation. Typical components of multi layered polyolefin coating are schematically shown in Figure 9.1.

The main selection criteria for polyolefin coatings is the operating temperature. Polyethylene is subject to a maximum operating temperatures of between 65 °C and

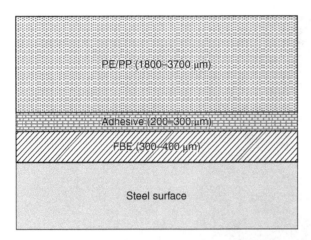

Figure 9.1 Schematic presentation of three-layer polyolefin coatings.

Figure 9.2 Completely disbonded three-layer PP coatings removed from a gas pipeline (see colour plate section).

75 °C while polypropylene can be used for maximum temperatures of between 110 °C and 140 °C, depending on its formulation.

An important fact to note about multi-layer polyolefin coatings is that their maximum temperature is the lowest maximum temperature resistance of any of its individual layer coatings. For example, in a three-layer PP coating, if the top polypropylene coating has a maximum temperature resistance of 120 °C with the first layer FBE coating only rated to 90 °C, then the maximum temperature resistance of the three-layer PP coating will be 90 °C. Failure to note this could lead to detachment of the FBE layer and hence failure of the three-layer PP coating, an example of which is shown in Figure 9.2.

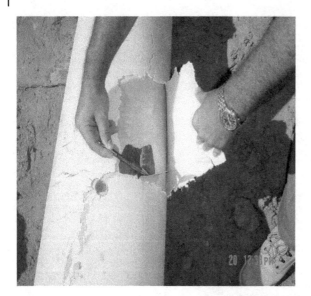

Figure 9.3 Cracking and disbondment of the three-layer PP coating due to thermo-oxidative degradation. (see colour plate section).

Three-layer polyolefin coatings are particularly suitable and popular for use on offshore pipelines. Their combination of corrosion mitigation capability, impact resistance, and some insulation is ideal for the offshore environment. With three-layer PP coatings, however, care is required if they are being used onshore. Failure has been reported in three-layer PP coatings used on gas lines, with a maximum operating temperature of 110 °C, buried in the desert in the Persian Gulf. Due to continuous movement of sand in the desert, parts of the line become exposed to the strong ultraviolet radiation. Under these conditions, these coatings are prone to thermo-oxidative degradation and high residual stresses in the PP. Cracking and coating disbondment have been found to occur due to high residual stress concentration and adhesion loss after moisture interaction or thermo-oxidative degradation of the FBE primer [2]. An example is shown in Figure 9.3. The top detached layer in the figure is the polypropylene. The green coating below it is the FBE and the black section is the exposed steel pipeline.

The German standard for polypropylene coatings of steel pipes shows that increasing the operating temperature of PP-coated pipes from 60 to 100 °C reduces the minimum expected service life of these coatings from 50 years to only 8 years [3] (Table 9.1).

9.1.3 Field Joint Coatings

Coated pipelines are taken to the installation sites with a small section at either end left uncoated. These 'weld joints' are coated after the pipeline sections are welded together. In some cases, the field joint coating can be the same as the 'factory or shop-applied' coating, for example, pipelines coated in FBE or liquid epoxy. In other cases, a different type of coating may be used on the field joints. The field joint coating, however, must be compatible with the main line coating.

Table 9.1 Typical relationship between the operating temperature and the minimum expected service life of polypropylene coatings for steel pipelines.

Operating temperature (°C)	Minimum expected service life (years)
23	50
60	50
80	30
90	15
100	8

Figure 9.4 Failure of a flame-sprayed PP field joint coating on a three-layer PP-coated gas pipeline. (see colour plate section).

For several reasons, weld joint coating are areas that are the weakest points in the pipeline. Welding induces stresses in the steel and, if not properly post-weld heat-treated, these stresses remain as residual stresses, raising the potential for subsequent corrosion damage and cracking of coated areas later. The coating of the welded joints takes place in the field conditions. It is, therefore, unlikely that the coating application, and its QC/QA, will be as much under control as in the factory and being conducted under time pressures to maintain pipe-laying schedules. Often, the first place where the pipeline coating system fails is at the field joints. This has particularly been observed in field joint coatings for PP-coated pipelines [4]. An example is shown in Figure 9.4, where the flame-sprayed PP field joint coating on a three-layer PP-coated gas pipeline has cracked and detached from the steel substrate.

There have been continuous improvements in field joint coating technology for polypropylene-coated pipelines. One such example is application of hot PP tape, as shown in Figure 9.5.

One of the most popular types of coating system used for field joints of pipeline with any type of shop coating is heat-shrink sleeves. These are available in either PE or PP and have an epoxy backing. They are placed loosely around the weld joint. They are then heated, either manually with flame torches or by using various heating

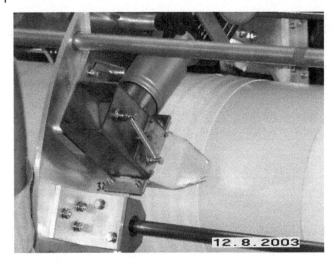

Figure 9.5 Application of hot PP tape to the weld joint of three-layer PP-coated pipeline. (see colour plate section).

(a)   (b)

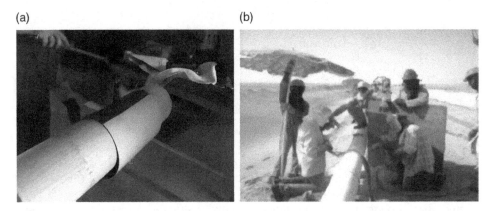

Figure 9.6 (a) Heat shrink sleeve coating on a weld joint in the desert; (b) application. (see colour plate section).

instruments. During heating, the sleeve shrinks and fits tightly around the joint (Figures 9.6a and 9.6b).

Polyurethane coatings can also be used for field joint coatings and they cover a wide range of temperature resistance.

9.2 Internal Coating and Lining

Internal coating and lining can be carried out on any type of pipeline or downhole tubular: oil, gas, or water. For oil and water, the main purpose of the coating is to protect from corrosion. In gas pipelines, it can have an additional purpose which is to improve the flow of gas. The most popular internal pipeline coating systems are FBE and liquid epoxy. Linings are further discussed in Chapter 15.

Figure 9.7 Field insertion of HDPE liner in a steel flowline. (see colour plate section).

Internal lining normally refers to insertion of a solid non-metallic pipe, such as high-density polyethylene (HDPE), into a steel pipe. This can either be done at the factory before shipping the pipe for installation or on site for rehabilitation purposes (Figure 9.7) by swaging a section of the liner through the solid pipe. HDPE liners are particularly suitable for water service: water supply, wastewater, water injection, etc. It complements a steel pipeline's mechanical strength with HDPE's corrosion resistance.

9.2.1 Plastic-Coated Tubular (PCT)

Several plastic coating systems are available which are used to coat internal surfaces of well tubulars. These are mostly based on a liquid-phenolic or powder epoxy novolac (a type of phenolic resin) compounds. Industry experience of their use is mixed and their use is primarily for water injection systems. Plastic-coated tubular (PCT) is generally regarded as a temporary fix rather than a permanent solution to corrosion protection. However, as explained in Chapter 7, PCT also offers drag reduction and other benefits, including scaling reduction and a possible solution for microbial corrosion.

The issues related to PCT are its low resistance to both wireline damage and well stimulation chemicals and also potential coating collapse in the event of rapid decompression. There is also a risk of possible crevice corrosion set in places where the coating is damaged.

As a result of poor mechanical robustness, the use of PCT in injection service is generally subject to the same corrosion limitations as described for bare CLAS referred to in Chapter 7. There are only a limited number of circumstances in which PCT is known to offer superiority and can be considered – these are as follows:

1) when handling non corrosive or adequately and reliably treated fluids;
2) when there is no intervention work or extensive use of acidic chemicals.

It should be noted that in corrosive service, PCT will ensure much better final mechanical integrity than plain carbon steel, and hence is likely to be easier to retrieve.

9.2.2 Glass Reinforced Epoxy (GRE) Lined CLAS Tubing

A non-metallic lining system has been deployed successfully in a number of well completion over the years. The system entails insertion of a fibreglass reinforced plastic (FRP) or glass reinforced epoxy (GRE) tube into carbon steel tubing host followed by the injection of a special grout into the annular space between the liner and the tubing.

The system maximises the performance and reliability of carbon steel tubing combining the high strength of the host pipe with the inert properties of the liner to provide an effective corrosion barrier. A special elastomeric ring in placed in each joint, hence providing a continuous barrier inside each pipe.

GRE lined tubulars have been used extensively for water injection applications when water quality is not continually controlled and in some instances for gas injection (including CO_2 sequestration), oil and gas production.

The use of GRE lined C-steel is becoming more popular in the industry, especially for new projects. However, again unless the produced water is suitably cleaned up to minimise the presence of any residual oil, care should be taken when selecting a lining for produced water injection, as most may have limited hydrocarbon service capabilities.

9.2.3 Internal Coating of Tanks and Vessels

There are many different types of coating system available for internal coating of tanks and vessels. The selection depends on a number of factors: operating temperature and pressure, corrosivity, design life, presence or absence of cathodic protection. Various types of epoxy coatings are the most popular choice. Other choices include polyurethane, glass reinforced vinyl ester, polyester, thermally sprayed aluminium, and ceramic coatings (for high temperature).

Non-metallic linings, such as high density polyethylene (HDPE) or GRE, can also be used. For very high corrosion resistance, often coupled with high temperature application, for instance, in pressure vessels, such as gas separators, thermally sprayed aluminium, and cladding with corrosion-resistant alloys (CRA) can also be used.

9.3 External Painting of Structures

The main deciding factors for selection of paint systems for external protection – cf. fabric maintenance – are normally the exposure environment, expected life, and cost. Aesthetic factors also often are part of the selection criteria. A typical coating system could be a zinc-rich or an epoxy primer with a second coat of high-build epoxy and a very thin top coat of polyurethane for improved resistance to corrosion, impact, and UV radiation.

9.3.1 Offshore Structures

Selecting the right coating for offshore structures is critical as recoating and coating rehabilitation are expensive. As a minimum, a multi-epoxy coating with anti-fouling reagents, or an extra anti-fouling coat are required. For splash zones, higher spec coatings such as glass reinforced vinyl ester or polysiloxane are recommended.

Table 9.2 A summary guide to coating applications.

Application	Corrosion barrier	Most widely used coating	Key attributes	Shortfalls
Pipelines	External	FBE	• Corrosion protection • Can use on field joints too • Long track record • High temp application	• Impact damage • UV degradation
	Internal	HDPE	• In-situ application • Mechanical strength	• Requires venting • Temp limitation
		Epoxy coating	• Corrosion resistance • Improved flow • Long track record	• QA/QC critical • Small holidays can cause leakage • Disbonding can cause blockage
Tubular	External (mainly surface casings/ conductors)	CP – coating not normally used here		
	Internal	GRE liners	• Corrosion control	• Temp limitation
		Epoxy coatings	• Wear prevention • Hydraulic Improvements • Prevention of scale deposits	• Temp limitation • Mechanical damage
Drill pipe	Internal	Epoxy coating		
Vessels	Internal	Epoxy coating	• Corrosion resistance • High temp application • Ease of application	Impact damage
		Vinyl ester	• Toughness • High temp application • Glass reinforced	QA/QC very important
		Thermally sprayed Aluminium	• Corrosion resistance • Toughness • High temp application	QA/QC very important Cost
Structures	External offshore	Polysiloxane	• Corrosion resistance • Toughness	Cost
		Multi coats of epoxy	Corrosion resistance	Fouling
	External splash zones	Glass flake reinforced coatings	• Corrosion resistance • UV resistance • Toughness	

9.4   Summary

Coating systems offer the most widely used means of affording corrosion protection for external surfaces of pipelines and structures. In addition, coatings and linings provide a corrosion barrier on internal surfaces to allow transportation of corrosive fluids. The latter takes advantage of the strength and economy of carbon and low alloy steel pipe with a degree of corrosion resistance.

This chapter was not intended to offer an exhaustive reference to all coating and lining systems used in hydrocarbon production systems, but rather aimed to provide a practical summary approach to the use of such systems. Table 9.2 summarises the use of coating and lining systems with key attributes and shortfalls that can be used as a reference for practical purposes.

References

1 A N Moosavi, Advances in field joint coatings for underground pipelines. NACE Annual Congress, Paper No. 00754, 2000.
2 A N Moosavi, BTA Chang and K Morsi, Failure analysis of three layer polypropylene pipeline Coatings, NACE Annual Congress, Paper No. 10002, 2010.
3 Deutsches Institut für Normung, Polypropylene coatings for steel pipes. DIN No. 30678, October 1992.
4 A N Moosavi, Hidden problems with three-layer polypropylene pipeline coatings. NACE Annual Corrosion Conference, Paper No. 06057, 2006.

Bibliography

API, Recommended practices for unprimed internal fusion bonded epoxy coating of line pipe. API 5L7, 1988.
API. External fusion bounded epoxy coating of line pipe. API 5L9, 2001.
DNV, Pipeline field joint coating and field repair of line pipe coating. DNV-RP-F102, 2011.
ISO, External coatings for buried and submerged pipelines used in pipeline transportation. ISO 21809, 2011.
Kehr, J.A. (2003). *Fusion-Bonded Epoxy (FBE): A Foundation for Pipeline Corrosion Protection*. NACE.
NACE, Field-applied heat shrink sleeves for pipelines: application, performance and quality control. NACE RP0303, 2003.
NACE, Application, performance and quality control of plant-applied fusion bonded epoxy external pipe coating. NACE RP0394, 2013.
NORSOK, Surface preparation and protective coating. NORSOK Standard M-501, 2012.

10

Corrosion Trending

Corrosion trending covers all aspects of monitoring and inspection and their integration into an effective strategy essential to fit-for-service (FFS) facilities. This involves having, at all times, full understanding of internal and external exposure conditions, and a sound knowledge of the physical condition of each component within an asset. The term corrosion trending is used in this chapter to identify the integration of monitoring and inspection in detecting and quantifying the presence of a corrosion threat and its timely and sustained mitigation. While the current condition of the equipment can be determined using inspection techniques, corrosion monitoring enables information to be gathered on the corrosivity of the environments that the equipment is exposed to. Inspection involves quantifying the equipment wall thicknesses and identifying defects using a range of available techniques. To complement inspection methods a leading indicator is needed; this is achieved in part by what corrosion monitoring provides by means of a range of techniques.

Apart from corrosion monitoring and inspection, a holistic approach to ensuring system well-being requires two additional data sets: (i) analysis of process streams; and (ii) assessment of operational history. The former is through monitoring and analysis of key process variables that influence corrosivity, such as temperature, pressure, production rate, fluid chemistry, corrosion product concentration, and gases. The latter provides information on present and predicted corrosion rate through assessment of operational records (process changes), failure analyses, and inspection records. However, these two themes are considered a routine part of operation and therefore not detailed in the present chapter. The four elements of inspection, monitoring, analysis of process streams and assessment of operational history are integral to corrosion trending and form a key component of a corrosion management strategy that is discussed in Chapter 18.

The present chapter outlines the key aspects of corrosion monitoring and inspection that a corrosion engineer needs to consider when building corrosion management programmes. It does not discuss specific techniques in detail and refers to other available publications that cover the respective methods. The chapter provides

Corrosion and Materials in Hydrocarbon Production: A Compendium of Operational and Engineering Aspects, First Edition. Bijan Kermani and Don Harrop.
© 2019 John Wiley & Sons Ltd. This Work is a co-publication between John Wiley & Sons Ltd and ASME Press.

guidelines outlining the advantages and limitations of each method as well as looking to the future and the blurring of corrosion monitoring with traditional inspection technologies. In this chapter, the term corrosion monitoring embraces both corrosion and erosion monitoring.

10.1 The Purpose of Corrosion Trending

To minimise safety, environmental, and business risks, it is essential that the fluids remain contained inside the equipment. To do this and maximise the uptime and lifetime (and hence profit) of the facilities, they must be maintained in a condition appropriate for the service required. This condition is often described as FFS (fit for service). Inspection involves quantifying the equipment wall thicknesses and identifying defects, such as cracks, pits, or bulges using one or more of the available techniques. Inspection is the most accurate way to determine current equipment condition but it has the obvious disadvantage that any damage that is detected has already occurred. Inspection is therefore a lagging indicator.

Therefore, to complement inspection methods, a leading indicator is needed; something that will identify that degradation is occurring and provides sufficient warning so that an intervention can be implemented well in advance of the problem becoming critical. In practice, a true leading indicator is difficult to obtain but this is what corrosion monitoring strives to do and there is a range of techniques available to detect specific corrosion mechanisms and corrosion rates. Corrosion monitoring techniques have their limitations in terms of both precision and accuracy.

A combination of inspection and corrosion monitoring built into corrosion trending is a core element of a sound corrosion management programme. The use of either one individually exposes the equipment user to unnecessary risks and possibly the expense of less than fully effective mitigation programmes.

A good corrosion management programme can be thought of as a continuous process, the aim of which is to keep the equipment fit for service (FFS) for the lifetime determined by the facility owners: this is discussed in more detail in Chapter 18. One challenge for the corrosion engineer is that the required facility's life can be a moving target which depends on financial, political, and technological considerations. To determine if equipment will be FFS for a given lifetime, the current condition (inspection) and corrosion rate (monitoring) must be known and continually re-checked.

The remaining life of a piece of equipment is a function of the wall thickness, the corrosion rate and the overall effectiveness of all the corrosion mitigation programmes. In practice, the latter is typically evaluated by calculating or estimating the effectiveness of each individual mitigation programme (e.g. corrosion inhibition, biociding, coatings, cathodic protection, etc.) and using the lowest remaining life obtained. Where corrosion-resistant materials have been chosen, the general corrosion rate should be very low or zero. However, thought must also be given to potential pitting and localised corrosion or non-routine operation mechanisms. For example, a vessel may be clad internally with AISI 316 stainless steel, in which case the corrosion rate due to carbon dioxide will effectively be zero. However, if it is frequently cleaned out using untreated sea water, the pitting corrosion rate may be significant, albeit difficult to estimate.

10.2 Corrosion Monitoring

Historically corrosion monitoring was exclusively associated with the measurement of corrosion rates. However, it is important to recognise that it also includes the measurement of the performance of corrosion barriers, for example, the availability of corrosion inhibitors, the condition of coatings, or the electrical potential of equipment under cathodic protection control.

10.2.1 Corrosion Rate Monitoring

It is possible to monitor the corrosion rate by inspection only where any wall loss is measured at periodic intervals and from this an estimate of remaining life is made. Where this is done, care must be taken to understand the limits of the inspection technique in terms of sensitivity and detection probability, as well as the frequency of repeat inspections. However, as noted above, this is a reactive approach and once corrosion has occurred, any wall loss cannot easily be replaced. Moreover, an inspection measurement only provides the overall corrosion loss since the last inspection. It cannot identify if the corrosion rate is steady, decreasing, or increasing unless the inspections are repeated at very short intervals: this is normally cost-prohibitive. In reality, the majority of inspections are carried out on a one-to-five-year frequency.

Once a piece of equipment has been built, the materials of construction are fixed and so corrosion monitoring is essentially the study of changes in the corrosivity of the electrolyte towards the material of the equipment. The overall aim of corrosion monitoring is, therefore, to have a leading indicator of the potential for corrosion to equipment before significant damage occurs and allow mitigation methods to be employed.

Typically, the external and internal corrosivity of a system will vary with time for many reasons such as:

- changes in environmental conditions;
- breakdown of coatings;
- temperature (e.g. day and night) and/or pressure changes;
- fluid chemistry changes;
- flow regime and/or velocity changes;
- corrosion inhibitor changes (including concentration variations).

It is, therefore, very valuable to have an estimate of the corrosion rate on a much shorter timescale than the inspection interval and this is the objective of a corrosion monitoring programme. For external corrosion, this is typically done using visual or other inspection techniques due to the relative ease of access to the external surface, although significant exceptions to this are buried, subsea or insulated equipment. However, the external corrosion rates of these systems are still almost always performed using inspection techniques such as radiography or in-line-inspection. For internal corrosion monitoring, a broader range of techniques is used, as discussed below.

In broad terms, corrosion rate monitoring is the measurement of a representative corrosion rate for a given piece of equipment exposed to a corrosive service. There are several techniques that can be used either as standalone or in concert with each other.

Ideally corrosion monitoring is designed to provide real-time feedback on the corrosion control process present. It is important to remember that any given monitoring technique will have limited accuracy and sensitivity and should be chosen to provide the appropriate information.

The traditional methods for corrosion monitoring in the oil and gas industry are:

1) weight loss coupons;
2) electrical resistance (ER) probes;
3) electrochemical monitoring.

A large body of information is available on each technique [1, 2] and only a brief description of each is given here. The advantages and limitations of the three traditionally most commonly used corrosion monitoring techniques are summarised in Table 10.1.

10.2.2 Weight Loss Coupons

Weight loss coupons are the most commonly used method for determining corrosion rates (often referred to simply as 'coupons') due to their simplicity of use and relatively low cost. This involves direct exposure to the potentially corrosive environment of typically duplicate metal coupons mounted on a retrievable holder inserted into a pipeline, facility, etc. The coupons are constructed ideally of the same metal as the equipment of concern. However, in practice, this can be difficult to achieve so a generic steel with properties as close as possible to the actual equipment is used. Differences between weights prior and subsequent to exposure are then translated into a corrosion rate by taking into account exposure period. The basis of the technique is easy to understand and it does not require complex theoretical knowledge or analysis techniques to obtain data.

10.2.3 Electrical Resistance (ER) Probes

ER probes are the second most widely used technique and, as the name suggests, this technique is based on the measurement of the electrical resistance of a probe in a corrosive environment. Similar to weight loss coupons, the sensing element of an ER probe is constructed from a strip or wire ideally of the same metal as the equipment of concern, but as with coupons more commonly from a generic steel. Suppliers of coupons and probes keep a range of such materials and can be consulted for help on the best material to use. As with coupons, the probes can be designed to be flush-mounted with the surface of the equipment or inserted into the bulk of the corrosive environment. This is an important consideration, for example, where flow conditions (e.g. turbulent versus stratified flow) and the presence of solids have a critical bearing on corrosion rate and morphology (e.g. pitting). as it is critical not to impair the passage of a cleaning or inspection tool through a pipeline.

Over the last few decades several manufacturers have attempted to build multiple arrays of sensors into a single tool to overcome the spot reading and enable wider area coverage. These include techniques, such as the field signature method (FSM) and ring pair corrosion monitoring (RPCM) tools. These techniques have received mixed reviews, especially for subsea monitoring. Because of this and their very high unit and installation cost, they have become less popular and the FSM technique is no longer sold in a subsea version.

Table 10.1 Advantages and limitations of corrosion monitoring techniques.

Corrosion monitoring techniques	Advantages	Limitations
Weight loss coupons	Simplicity and low cost Can be used in any environment Corrosion product available for analysis Ease of visual examination and option to undertake qualitative and quantitative surface analyses Direct measure of metal loss Possibility of detecting erosion damage	Non-instantaneous Require insertion into the fluid Medium-term response time Cannot be automated Corrosion rate is averaged over exposure period (i.e. lagging measure of corrosion rate) Installation and retrieval can be time-consuming
ER probes	Will operate in any environment Provide 'real-time' measurements that show how corrosion rate changes over time and with changes in conditions Relatively simple in operation and interpretation Possibility of detecting erosion rates by using an element made from a non-corrosive metal (typically 316 stainless steel) Corrosion or erosion rate can be measured at any frequency The rate is delivered as an electrical signal which allows for online monitoring from any location	Not designed to monitor localised corrosion A higher unit cost option than weight loss coupons Do not give instantaneous corrosion rates They can be fouled by deposits in certain environments, specifically, in H_2S environments Judgement needed to balance the sensitivity of the probe with its lifetime Sensitive to thermal changes
Electrochemical monitoring (LPR probes)	As per ER probes Rapid response to process change Possible to detect localised corrosion It provides 'instantaneous' corrosion rate data and hence fast response to system upsets As such, it is a valuable means of activating alarm and countermeasure systems Probes are usually more rugged and less expensive than ER types	Will provide reliable data only in electrically conductive fluids The probes are prone to error in systems with high levels of entrained debris In sulphide-containing environments under- and over-estimates of corrosion rates may be indicated

10.2.4 Electrochemical Methods

Electrochemical monitoring comprises a family of monitoring methods that are based on the understanding that corrosion is an electrochemical phenomenon, i.e. the simultaneous combination of electrical and chemical processes. Over the years many measurement techniques have been developed and combined with numerous specific probe designs for the investigation of specific corrosion mechanisms. The study of these techniques and methods is a distinctive discipline beyond the scope of this chapter and is described elsewhere [3, 4].

The techniques most commonly used in the oilfield are corrosion potential, linear polarisation resistance (LPR), wide potential scans (Tafel plots), alternating current impedance (AC impedance also known as electrochemical impedance [ECI]), and electrochemical noise (ECN).

For reasons outlined in Table 10.1, electrochemical techniques are used in oil and gas production environments in a limited way [5], such as relatively clean water systems (sea water and some produced waters) or in laboratory studies to estimate the potential corrosion rates and evaluate corrosion inhibitor effectiveness. When they are used, operators often use the trend in corrosion rates rather than absolute values.

10.2.5 Locating Internal Corrosion Monitoring Devices

The number of monitoring locations, their location, and orientation on the equipment as well as the choice between intrusive and flush-mounted probe designs are important decisions for the corrosion engineer which can depend on many factors including:

- criticality of the equipment;
- physical access;
- expected corrosion mechanism;
- confirmation that certain corrosion mechanisms are not active;
- budget.

It should be borne in mind that it is the corrosion monitoring device that experiences a particular corrosion rate at a specific location. The equipment itself may or may not be experiencing exactly the same rate or morphology of attack. Typically, there would be more than one monitoring location and rates should be verified with inspection data. The choice of monitoring location(s) is therefore one of the most important decisions a corrosion engineer must make and the phrase 'location, location, location' from the real estate business is therefore very appropriate for corrosion monitoring.

Often the best location is determined using the experience and judgement of the corrosion engineer or team. This is usually done through a combination of the use of predictive corrosion models (discussed in Chapter 5) and an understanding of the flow characteristics of the fluid (e.g. stratified, slug, or annular flow). This works well for pipelines but for short-run piping systems, it can be more difficult.

In recent years, advances in computational fluid dynamics (CFD) have begun to help with identifying the best locations for monitoring in piping systems, especially potential corrosion hotspots due to changing flow conditions (e.g. dead legs, intrusions, bends, and changes in elevation). This can have increased significance where solids are present with the attendant potential threats of erosion, erosion-corrosion, and corrosion under deposits. By way of example, Figure 10.1 shows the result of a CFD analysis which identifies the location of maximum erosion rate in a piping run from a dry gas wellhead.

Even with the benefit of the analysis in Figure 10.1, it may not be possible to determine a single, precise location and so it may require the use of several probes to maximise the probability of detecting damage. CFD technology can also be used to determine the optimum locations for inspection.

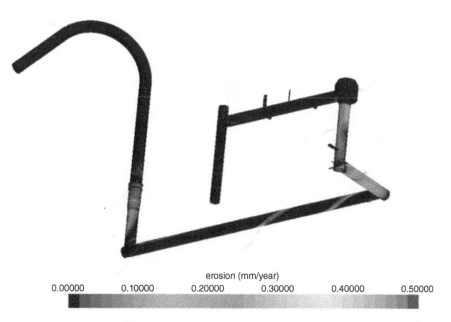

erosion (mm/year)

| 0.00000 | 0.10000 | 0.20000 | 0.30000 | 0.40000 | 0.50000 |

Figure 10.1 CFD model showing relative erosion rates in well-head piping. Red colour denotes the highest rate. (see colour plate section).

10.2.6 Erosion Rate Monitoring

Erosion can occur at extremely fast rates that are often measured in minutes and hours rather than years. Moreover, erosion is rarely a continuous process and comes in intermittent bursts. In extreme cases, erosion has been observed to cut through 25 mm (1 in.) pipework in 1 hour, equivalent to >200 m year^{-1}). Erosion monitoring, therefore, presents a significant challenge which can only be addressed by continuous monitoring, such as use of ER probes.

Another option which is very common is the use of sand detection probes, often referred to as acoustic monitors. These probes are essentially externally mounted microphones that listen for the sound created when solid particles, such as sand, impact onto piping. They have proven to be very successful in gas systems which are generally the most erosive due to high gas velocities. They have been less successful in liquid systems as the liquid can cushion the impact and reduces the signal-to-noise ratio. There have been attempts to use these probes to quantify the volume of sand produced as this is an important variable for predicting erosion rates. However, precision of initial calibration and need for subsequent recalibration if notable erosion damage has occurred where the monitor is located mean significant caution needs to be exercised if used in this mode.

Fixed ultrasonic probes also find use for erosion rate monitoring. For erosion monitoring these probes are usually tied directly into the operating software and set to alarm at particular values to allow control room staff to determine if an intervention is required.

10.2.7 Access Fittings

Assessment of the degree of corrosion damage by means of corrosion monitoring requires access to the location of the interface between the material of construction and the prevailing media. For atmospheric corrosion (usually referred to as external corrosion), access to the electrolyte is usually relatively simple. For internal corrosion, access to the fluids can only be achieved by the insertion of a probe into the process stream. There are several ways to do this, including the use of specially designed flanges and piping sections. However, the most common method in the oil and gas industry involves the use of an access fitting which is welded onto the equipment. These fittings provide an opening into the fluids through which a monitoring device can be inserted and retrieved. The most common fitting has a 2-in. (50 mm) opening through it and can be purchased to contain pressures as high as 42 MPa (6000 psig).

Figure 10.2 shows a typical access fitting 'kit' which includes the fitting (prepared for welding onto a pipe), a solid plug which is installed until a probe is actually used, a cap, and a pressure gauge (to indicate potential leaks). Lower pressure fittings (1 MPa, 150 psig) are also available with a 25 mm (1 in.) diameter opening. In principle, a fitting can be installed on-line but normally requires the system to be shut down for safety reasons. Probes and coupons are designed to be installed and extracted on-line.

The fittings can be attached onto the equipment wherever there is a suitable space, taking into account not only the size of the access fitting but also having sufficient access to safely use a probe/coupons retrieval tool. For piping, while default locations are commonly at the 12 and 6 o'clock positions, consideration needs to be given to ease of access for probe/coupons retrieval, the composition of internal fluids and the flow conditions as they can influence the type and wall location of corrosion likely to occur

Figure 10.2 Access fitting: kit and example installation. (see colour plate section).

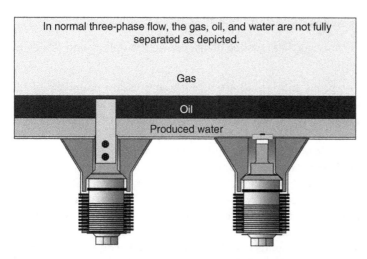

In normal three-phase flow, the gas, oil, and water are not fully separated as depicted.

Gas

Oil

Produced water

Figure 10.3 Schematic of intrusive and flush-mounted coupons.

(e.g. bottom of line; top of line; erosion corrosion; under deposit attack). Similar considerations apply to the choice of intrusive or flush-mounted probes and coupons as shown in Figure 10.3. Once a suitable monitoring location has been found, it is common to install two or more access fittings to allow the output from different monitoring techniques to be obtained and compared.

10.2.8   Cost Considerations

Clearly there is a cost to installing and running corrosion monitoring programmes and this needs to be balanced against the value that they will provide. Installation during the construction of a facility is always the most cost-effective method but in many cases the locations determined during design are often not ideal and a compromise or retrospective installation is required. Monitoring equipment costs are relatively inexpensive but the cost of welding access fittings in place may be a factor of 10 times this. This factor can be even higher for locations that are difficult to access.

For ER probes, the ideal situation is to have them hard-wired or wirelessly connected into the equipment control a system which, again, is best done during design and construction. In many cases retrofitting instrumented corrosion monitoring into existing software and wireless networks is often cost-prohibitive and may raise compatibility issues.

There can be a significant operating cost to manage coupons and probes which obviously depends on the size of the programme. Insertion and retrieval of coupons and probes must be done by specially trained personnel. Analysis of coupons requires laboratory facilities and analysis of data requires appropriate training. These contribute to the cost of the programme and so the value of the data must be carefully considered. It should never be considered as simply nice to have. If the data are not actively used and acted upon, it begs the question of why invest in the expense and trouble of installing corrosion monitoring facilities at all.

10.2.9 Safety Considerations

There are several safety concerns that must be managed when installing and retrieving coupons and probes:

1) To get the coupon or probe into an access fitting, the pressure boundary of the equipment is intentionally breached. There is, therefore, an increased risk during these operations for a leak to occur which could range from a minor release to a catastrophic event.
2) Many of the retrieval tools used are heavy and require two people to operate. Care must be taken to handle the tools properly to avoid injury. In addition, the external caps that act to protect the access fitting and provide a second pressure barrier have on occasion become unscrewed and fallen to the ground, risking injury to those nearby. The industry has been able to address these risks to some extent with better designs but the risks cannot be completely eliminated.
3) Many oil and gas fluids contain solids and debris which can become trapped in the access fitting. During retrieval these solids can result in galling of the threads which makes retrieval impossible. In some cases, this has led to the retrieval tool having to be left attached to the access fitting which introduces obstacle and leak hazards. Removal of such a tool typically requires a full shut-down of the equipment and the damaged access fitting must be repaired.

To manage these risks, specially trained technicians must carry out these tasks. However, some operators or plant managers have made the decision not to allow on-line retrievals and these are only done during equipment shut-down windows.

10.3 Corrosion Barrier Monitoring

In oil and gas production, unmitigated corrosion rates can be as high as $25\,\mathrm{mm\,year^{-1}}$ ($1\,\mathrm{in.\,year^{-1}}$). To reduce corrosion rates to an acceptable level, corrosion engineers use a variety of mitigation methods known as barriers. These fall into two broad categories as follows:

1) *Passive barriers*: these are barriers which require little or no active management during the lifetime of the equipment. The most common ones are the use of a material that is resistant to corrosion in the specified fluid or additional wall thickness that is consumed by corrosion (known as corrosion allowance).
2) *Active barriers*: These are barriers that require active management by corrosion engineers. This can range from infrequent, visual inspection to monitor the condition of paint coatings to daily adjustment of corrosion inhibitor injection pumps.

It should never be assumed that because a barrier has been installed, it is always working as designed. Where active barriers are employed, it is essential that their performance is monitored and this is known as corrosion barrier monitoring. It is important, therefore, to recognise that the success of a corrosion monitoring programme is not just about directly monitoring the corrosion rate. Related contributory or circumstantial data and evidence bring important added context to the significance of point sources of measured corrosion rate.

A good corrosion management programme will have at least one barrier in place for each credible corrosion threat and each of these barriers should be monitored to ensure they are being applied as designed. It should be remembered that measured corrosion rates are the consequence of all the barriers in place. Examples of corrosion barrier monitoring include:

- electrode potentials of equipment under cathodic protection at key locations;
- the number of planned maintenance (cleaning) pig runs are being met;
- chemical availability (e.g. corrosion inhibitor, biocide, oxygen scavenger);
- flow rates where flow restrictions have been implemented;
- temperature monitoring where temperature limits have been specified;
- monitoring bacteria;
- oxygen monitoring.

Corrosion barrier monitoring parameters form an important part of an overall corrosion management programme, as discussed in Chapter 18.

10.4 Collection and Analysis of Real-Time Monitoring Data

Following data acquisition by corrosion monitoring and inspection, it is paramount that the data are stored, analysed, and interpreted. The volume of acquired data should be anticipated so that the corrosion trending programme can be optimised accordingly, taking into account this otherwise potentially laborious and time-consuming task. The data trending should be performed and the results presented preferably in a graphical format to ease the corrosion management programme (see Chapter 18).

Real-time transmission of corrosion data from electrically based monitoring (e.g. ER, LPR, oxygen probes) has been available for many years although it requires the costly installation of hard-wiring from the probe to a control centre. In recent years there have been significant advances in the availability and reliability of wireless communications. This has enabled corrosion monitoring data to be transmitted relatively inexpensively from remote locations. Some manufacturers of corrosion monitoring equipment now provide wireless systems that (relatively) easily integrate into existing control systems to allow real-time display of corrosion rate data. The software is often enhanced with features that send alert emails to the appropriate personnel when corrosion rates exceed pre-defined limits.

Many companies provide software that can take multiple data inputs and correlate them with the corrosion monitoring data. As an example, taking temperature, pressure, and flow rate data from a pipeline to estimate an unmitigated corrosion rate. Similarly, corrosion inhibitor injection rate data which determines the inhibitor concentration present can be collected to enable an on-going estimate of the inhibited corrosion rate as determined from laboratory or field corrosion test data compared to the corrosion rate from probes. These data are then presented in a corrosion dashboard which can be seen at any location around the world. Figure 10.4 shows a typical dashboard that displays real-time fluid flow rates, velocities, sand rates, and estimated corrosion rates.

It should be noted that reliance solely on measurement of a corrosion rate – it being either lagging (coupons) or 'real-time' (assuming the probe is functioning as designed

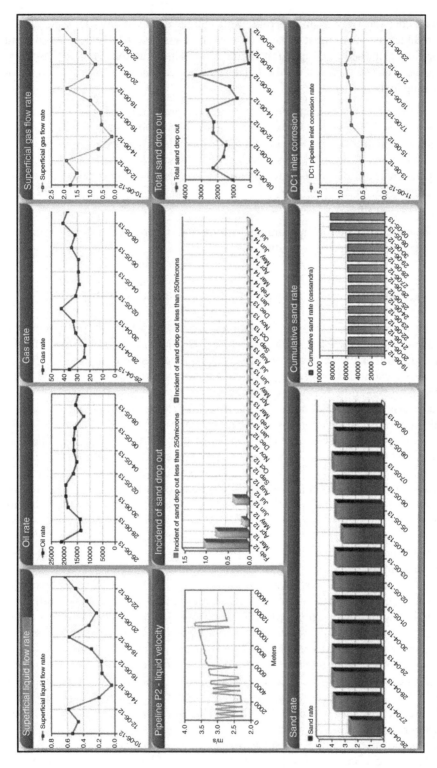

Figure 10.4 An example of a corrosion management dashboard. (see colour plate section).

so that a flat line is a true response) – can create a false sense of security (or alarm) when correlation with other related (albeit circumstantial) data can bring added and important clarity and assurance to the picture.

Typical dashboard data in Figure 10.4 are presented purely to show the various parameters that are readily and collectively measured and correlated to provide additional insight and clarity to the measured corrosion rate. However, the key point is that there is a large amount of potentially valuable data available – in the past, the value of its significance has not been appreciated or been able to be suitably captured – that greatly enhance understanding of the corrosion situation/threat present that corrosion monitoring alone may be under- or over-reflected. While artificial intelligence (AI) may offer a powerful tool to counter this human frailty, it comes down to proactive and consistent management with the right materials and tools in place to address the overall issue.

10.5 Downhole Corrosion Monitoring

While corrosion monitoring is a widely used method of corrosion management, by and large its use is limited to subsea, buried, and surface facilities. Prevailing conditions including high pressures and temperatures downhole, coupled with practical difficulties and the expense of placing monitoring instruments in the wells, have made downhole corrosion monitoring (DHCM) a rarely used venture. Usually DHCM is not economic, particularly if it involves intervention. Therefore, it is considered secondary to routine (basic) integrity checks. Generally, DHCM methods include fluid monitoring, workover inspections (downhole and recovered equipment), in-situ surface profiling calliper tools (mechanical and other), UT logs and other surveys, all of which are subjects beyond the scope of the present chapter.

Nevertheless, it is feasible to place corrosion coupons downhole via wirelines. These only apply to metal-loss corrosion, so are almost exclusively applicable to carbon and low alloy steels tubing and casing. These coupons can be retrieved during workovers and provide historical information regarding downhole corrosion trends and system corrosivity.

Some electrochemical monitoring methods, such as electrical resistance, electrochemical noise and impedance probers, have been tried on a limited scale, mainly in water injection wells. These have not yet gained wide acceptance and usage in the industry. They are also installed via wirelines but introduce added operational complexity due the need to be hard-wired back to the measurement instrumentation. Some of these probes are claimed to be usable in pressures up to 69 MPa (10 000 psi) and temperatures up to 150 °C. They use batteries which require to be changed periodically and normally less than three months.

Instead of corrosion monitoring, the focus in water injection wells has been on corrosion prevention by using non-metallic casing and tubing, or non-metallic liners as discussed in Chapters 7, 9, 14 and 15.

There are four corrosion monitoring/logging methods commonly used downhole: mechanical callipers; ultrasonic acoustic tools; electromagnetic tools; cameras.

Multi-fingered callipers are well-established tools that have been in common use for many years that provide reasonably accurate information regarding internal metal loss throughout the length of the tubing; but provide no data about external corrosion and are affected by scale build-up. However, when used in conjunction with other monitoring tools on the same wireline, they can provide additional valuable corrosion-related information, such as coverage and quality of cementing and depth of coverage of cathodic protection current (if cathodic protection is being applied to the casings).

Ultrasonic measurements yield excellent pipe thickness information and superior azimuthal resolution but these tools are not directly suited to use in gas wells. There is growing interest in the development and use of electromagnetic (EM) tools based on flux leakage and electromagnetic leakage (cf. MFL intelligent pig technique). They show good vertical and thickness resolution and detect internal and external metal loss. Cameras have also been used with some success for corrosion detection providing the wellbore is filled with gas or a clear liquid.

Due to their relatively high cost and disruption to well service, inspection tools tend to be used only every few years on candidate wells. Tool type and frequency tend to be on a case-by-case basis with a number of companies providing this specialist service.

In water injection wells, ensuring continual control of water quality, particularly oxygen content, is paramount. This can be achieved by placing oxygen monitoring systems at the wellhead to identify upset conditions.

10.6 Inspection Techniques

The majority of inspections are still carried out using well-established techniques [1] that have been available for many years, i.e. visual testing (VT), ultrasonic testing (UT), radiography testing (RT), magnetic particle testing (MT), and dye penetrant testing (PT).

Many of these techniques have been built into both internal and external tools, e.g. intelligent (smart) pigs using magnetic flux leakage or ultrasonic techniques, drones, and subsea remote operating vehicles (ROVs). A number of attributes are worthy of consideration as outlined in the following sections.

10.6.1 Equipment Portability

Many instruments are now much smaller to the point where they are truly portable and, in some cases, can be hand-held by a single person. As an example, a decade ago a 'portable' RT unit weighed about 20 kg (50 lbs.) and was typically mounted on a small tractor that could move along a pipeline. Today a single person can operate one comfortably, as typically shown in Figure 10.5.

Another important development is the increased use of remotely controlled crawlers and drones which can carry cameras to locations that are difficult or costly to access, such as subsea pipelines, flare stacks, and offshore platform jackets. Other applications include the use of infrared cameras to look for pipeline leaks or breaks in the thermal insulation. This equipment has a relatively low cost and allows certain inspections to be done while the equipment remains in service.

An important development in radiography is the widespread use of digital radiography which uses electronic detectors instead of traditional film plates. The resolution of the digital 'plates' provides very high-quality images with each pixel offering 250 μm

Figure 10.5 A one-person-operated, portable RT unit in use. (see colour plate section).

resolution. The high sensitivity also allows either lower strength radiation sources to be used or shorter exposure times.

In addition, modern data processing provides very fast data acquisition and analysis of images which allows the images to be seen in almost real time.

10.6.2 Visualising Inspection Data

The inspection of a given location will usually provide a great deal of information about any anomalies present. However, the data may be complex due to the geometry of the equipment and the shape of the anomalies. There is, generally, a limited ability to visualise such data in three dimensions (3D). A development that has made this much easier is the advent of 3D printing which builds a physical model of the equipment and any anomalies. Figures 10.6a and 10.6b show the inspection data of an anomaly in a pipeline and the corresponding 3D visualisation model printed using the data: the latter helps to make the inspection data more comprehensible.

10.7 Intelligent Pigging

Clearly there is a limit to the number of corrosion monitoring locations and physical extent of inspections that can practically be conducted, even before factoring in cost and logistics. This highlights the importance of first undertaking systematic corrosion risk assessments and operating risk-based inspection (RBI) programmes (both discussed further in Chapter 17).

However, for major pipeline systems and export pipelines, this may not be sufficient, even if directly accessible, in assessing and satisfactorily establishing their continuing

(a)

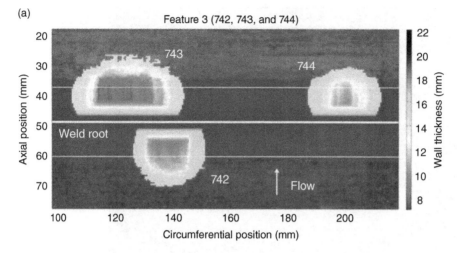

(b)

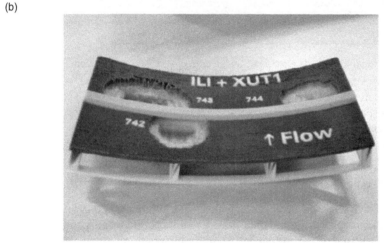

Figure 10.6 (a) Combined external and in-line inspection (ILI) UT data showing pipeline anomalies adjacent to a weld; and (b) a 3D-printed model of the pipeline anomalies. (see colour plate section).

FFS. Here periodic intelligent pigging is a key feature of a pipeline integrity management system (PIMS) with the frequency determined by the risk (often set not to exceed every five years).

Intelligent pigging requires the passage of an inspection tool through the internal bore of a pipeline with the ability to inspect the internal and external wall condition of a pipeline. It is disruptive to the operation of a pipeline, albeit for a short period, depending on the length of the line, and requires advanced preparation of a pipeline to ensure it has an acceptable level of cleanliness and physical condition. The latter is to avoid damage to the pig and for its smooth unhindered passage.

An intelligent pig commonly uses either magnetic-flux leakage (MFL) or ultrasonic (UT) on-board inspection technology via a circumferential array of sensors; but 'combo' MFL-UT tools are also available. The choice of inspection tool type will be on a case-by-case basis, depending on likely or known type of corrosion damage present, although a first inspection run is most often undertaken using MFL. More detail on the requirements for intelligent pig inspection is produced by the Pipeline Operators Forum [6]. MFL may be enhanced by the inclusion of the secondary technology of Direct Magnetic Response (DMR) to improve internal wall information and discrimination, especially for heavy wall pipe. There is also development of secondary eddy current in-line inspection (ILI) technology to enable early detection of what is termed shallow internal corrosion (SIC) an example of which is top-of-line corrosion (TLC). A geospatial tool will also be run separately and/or in combination with the inspection pig to measure distance, pig speed, rotation, internal geometry, and XYZ geographic coordinates, including that of welds.

UT needs the presence of a suitably conductive couplant between the inspection tool and the pipewall, typically treated sea water. MFL does not in principle require the presence of a couplant but the service provider will determine what pipeline environment best suits running their inspection tool. After satisfactory completion of an intelligent pig run, it is important to ensure couplant or selected running fluid is promptly and completely displaced, especially sea water.

Undertaking intelligent pigging is a highly specialised activity requiring particular expertise in its preparation, conduct and in analysing and interpreting the inspection data generated: something not solely carried and undertaken in-house. Also there is major capital investment in equipment development and hardware, equipment maintenance, software, etc. Needless to say, undertaking an intelligent pig run involves a significant cost, so budgeting as well as logistics should be accounted for in advance as part of a PIMS. There are number of companies that provide intelligent pigging services and these can readily be found via the internet.

10.8 Future Considerations

Inspection techniques can measure equipment wall thicknesses very accurately but historically they have required skilled technicians to make the measurements using portable equipment. The cost of this has meant that repeat inspections were undertaken at a frequency of one to five years. However, with improvements in technology, the use of permanently installed inspection equipment has blurred the boundary between what was traditionally referred to as inspection and monitoring and the use of inspection techniques as 'real-time' corrosion monitoring tools have become more common. This is a significant development since measurements of the actual equipment are very important. For example, the use of ultrasonic pulse-echo and pulsed eddy current (PEC) methods comprising a sensor matrix installed at fixed locations is gaining in popularity, including subsea deployment, and several manufacturers now provide systems for this purpose. These non-intrusive, highly sensitive technologies – typical absolute accuracy of 0.1 mm and repeatability of 2.5 μm – are able to work through solid external coatings (e.g. FBE, PE, 3LPP). They will likely become the preferred methods for corrosion monitoring going forward and offer the option to eliminate intrusive monitoring and the risks associated with it. At least one operator has adopted this approach and many others are considering it.

Guided-wave UT is increasingly being used to monitor long lengths of pipelines. Going forward, it is probable that these techniques will be used to provide close to 100%

coverage of equipment to provide real-time measurements at all locations. This will provide early warning that corrosion is occurring and allow engineers to follow up and intervene in a timely and appropriate response. This would be a key step towards intelligent equipment which tells us when any problem, not just corrosion, begins.

Corrosion trending to include monitoring and inspection programmes can generate large volumes of data which are often reviewed in isolation. In recent years there have been significant advances in data analytics (so-called 'Big Data'), artificial intelligence, and machine learning (also discussed in Chapter 18). These technologies can rapidly analyse vast quantities of structured (e.g. data) and unstructured (e.g. reports) information to provide insights that may have been missed.

Finally, engineers and technologists continue to find new and improved methods for monitoring and inspection. Perhaps one day corrosion may be eliminated but until then it is certain that better methodologies for monitoring and inspection will continue to appear.

10.9 Summary

Key aspects of corrosion monitoring and inspection built into an integrated corrosion trending programme that a corrosion engineer needs to consider when building corrosion management programmes are summarised in this chapter. While references have been made to different techniques with their advantages and limitations, no attempt has been made to discuss specific techniques in detail and reference has been made to other available publications that cover respective methods. Components of a holistic approach to corrosion trending and interrogation including corrosion monitoring, inspection, analysis of process stream and operational history assessment are outlined as a precursor to a corrosion management programme discussed in Chapter 18.

Aspects of safety, cost consideration, and location are briefly discussed. The concept of corrosion barrier monitoring is outlined to complement measures ensuring fit for service (FFS) facilities.

References

1 B Hedges, T Bieri, K. Sprague, and H Chen, A review of monitoring and inspection techniques for CO_2 and H_2S corrosion in oil and gas production facilities: location, location, location! NACE Annual Corrosion Conference, Paper No. 06120, San Diego, CA, 12–16 March 2006.

2 B Hedges, K Sprague, T Knox, and S Papavinasam, Monitoring and inspection techniques for corrosion in oil and gas production, NACE Annual Corrosion Conference, Paper No. 5503, Dallas, TX, 15–19 March 2015.

3 Papavinasam, S., Berke, N., and Brossia, S. (eds.) (2009). *Advances in Electrochemical Techniques for Corrosion Monitoring and Measurement*, STP 1506. ASTM International Publication.

4 NACE, Techniques for monitoring corrosion and related parameters in field applications. Technical Committee Report 3T199, 2012.

5 Papavinasam, S., Doiron, A., and Revie, R.W. (2012). Industry survey on techniques to monitor internal corrosion. *Materials Performance* 51 (2): 2–6.
6 Pipeline Operators Forum, Specifications and requirements for intelligent pig inspection of pipelines, 2009.

Bibliography

Al-Janabi, YT, Monitoring of downhole corrosion: an overview, SPE Saudi Arabia Section Technical Symposium and Exhibition, SPE-168065-MS, Al-Khobar, Saudi Arabia, 19–22 May, 2013.
Kermani, B. and Chevrot, T. (eds.) (2012). *Recommended Practice for Corrosion Management of Pipelines in Oil and Gas Production and Transportation*, Publication No. 64. The Institute of Materials, European Federation of Corrosion.

Specifications

DNV, Integrity management of submarine pipeline systems. DNV-RP-F116, 2009.
NACE, Techniques for monitoring corrosion and related parameters in field applications. NACE 3T199, 2012.
NACE, Field corrosion evaluation using metallic test specimen. NACE RP 0497, 1997.
NACE, Preparation, installation, analysis and interpretation of corrosion coupons in oilfield operations. NACE RP 0775, 2005.
NACE, Detection, testing and evaluation of microbiologically influenced corrosion (MIC) on external surfaces of buried pipelines. NACE TM 0106, 2016.
NACE, Field monitoring of bacterial growth in oil and gas systems. NACE TM 0194, 2014.

11

Microbiologically Influenced Corrosion (MIC)

A major corrosion threat in hydrocarbon production which continues to pose operational challenges and remains somewhat unimpeded is that caused by microbial activities. There is a perception that when the origin of damage is unclear, it is invariably associated with either microbiologically induced/influenced corrosion (MIC) or under-deposit corrosion (UDC). In both cases, severity of attack is high, understanding is limited, and mitigation methods are subject to discussion.

A basic definition of 'microbial corrosion' is given as the 'corrosion associated with the action of micro-organisms present in the corrosion system' [1]. The term MIC is used more recently. It highlights the fact that micro-organisms may not be the direct and unique cause of the corrosion mechanism but also may be enablers, enhancers, or inducers of corrosion damage in association with other corrosive contributors. In fact, without entering into semantic or mechanistic debates, the great virtue of this denomination is that it ends up as 'MIC', an easily expressive term. In this chapter only MIC or microbial corrosion terms are used, although the generic term of bio-corrosion might also be used.

MIC in the hydrocarbon production industry is usually perceived as an internal corrosion issue affecting primarily carbon and low alloy steel (CLAS) facilities. This is somewhat true because CLAS is presently the essential material of construction within the industry sector. However, micro-organisms may also be involved in the external corrosion of immersed and buried pipelines and structures as well as in the internal and external corrosion of stainless steels, among other materials used.

Nevertheless, this chapter focuses solely on internal MIC threats of CLAS – this is primarily due to the facts that:

1) The external corrosion of buried and immersed facilities, for both CLASs and stainless steels, is successfully prevented by cathodic protection (CP), in combination with organic coatings. The detrimental effect of micro-organisms is thus exceptional and mostly related to deficient protection.
2) There are very few reported cases of internal MIC issues on stainless steels. This is due to the fact that their resistance to the anaerobic MIC-prone environmental conditions of oil and gas facilities, including shut-down and stagnant situations, is generally associated with exposure to moderate temperatures which are not really critical to stainless steels.

Corrosion and Materials in Hydrocarbon Production: A Compendium of Operational and Engineering Aspects, First Edition. Bijan Kermani and Don Harrop.
© 2019 John Wiley & Sons Ltd. This Work is a co-publication between John Wiley & Sons Ltd and ASME Press.

This chapter addresses MIC from the point of view of an operating company or end user who may experience the damage caused by the threat and would need to prevent and subsequently resolve it. The chapter is not intended to be an exhaustive and detailed description of possible mechanistic aspects. It only concentrates on a few general and simple preliminary considerations and engineering solutions. The focus is placed primarily on dealing with the usual questions from engineering aspects, including the description, cause, influential elements, and how MIC occurs. It also refers to the means of mitigation and control, together with outlining effective monitoring methods to detect MIC.

11.1 Main Features

MIC, as an operational concern, is manifested by a deep localised corrosion typical examples of which are shown in Figures 11.1, and 11.2. It is evident that shallow pitting or light uniform corrosion can also be generated by micro-organisms alone, although such damage type never leads to any serious integrity issue; hence it is not a serious MIC concern in operations. Figure 11.3 is another typical example: numerous pit nuclei are observed all over the surface, but only the deep pit/grove in the middle may have potential integrity consequences.

As far as MIC morphology is concerned, the principal features include:

- There is no unique localised corrosion morphology to allow MIC to be diagnosed from an undisputable 'visual signature'.
- As shown in Figures 11.1–11.3, localised hemispherical damage is often observed. Unfortunately, very similar features are also experienced with the H_2S and CO_2 metal loss corrosion mechanisms as described in Chapter 5.
- 'Pits within pits' morphologies are sometimes observed, as reflected in Figure 11.2, showing that MIC progress may be intermittent.
- While being highly localised at their origin, MIC pits may coalesce over time to an extended corroded area. Figure 11.2 shows some areas where several 'pits' have started and subsequently joined together.
- Figure 11.3 shows a specific corrosion case recurrently observed in sea water injection networks, with a groove corrosion morphology at the bottom of some lines. Such cases are believed to be due to corrosion at the bottom of the line promoted by solid particle accumulation and enhanced because it is galvanically coupled with iron sulphide-covered surfaces eventually generated by bacterial activity. This is not classical MIC damage, but rather a microbial contribution to a pluri-factorial mechanism.

Common features of MIC damage include:

1) A localised corrosion phenomenon.
2) It is erroneous to conclude damage as MIC purely from visual observation of its morphology.
3) In most cases, it is preferable to diagnose the threat as a 'microbial contribution to corrosion' rather than an MIC or microbial corrosion.
4) Microbial corrosion, CO_2 corrosion and H_2S/CO_2 metal loss corrosion features can overlap under certain conditions and hence their differentiation needs caution.

Figure 11.1 Typical features of MIC type damage characteristic of localised morphologies. (see colour plate section).

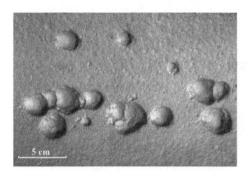

Figure 11.2 Localised hemispherical damage. (see colour plate section).

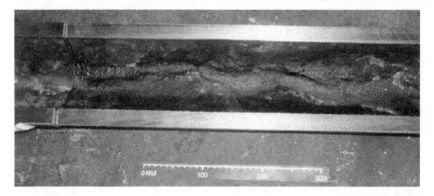

Figure 11.3 Groove corrosion promoted by microbial activity. *Source:* [2]. (see colour plate section).

11.2 The Primary Causes

A summary of key conditions necessary for the occurrence of internal MIC of CLAS equipment in most hydrocarbon production facilities is presented in Figure 11.4 and further summarised as follows:

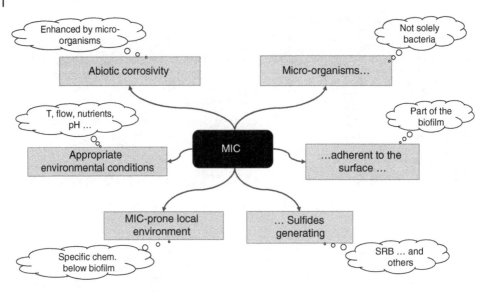

Figure 11.4 Key conditions necessary for the occurrence of MIC in hydrocarbon production streams.

- *Micro-organisms:* these are indeed necessary, by definition. Believing that MIC is only a bacterial corrosion should be avoided. Other micro-organisms than bacteria may also be involved in the process, directly or indirectly, as outlined below.
- *Adherence to the surface:* MIC is an electrochemical process requiring a charge transfer between the environment and the metal. An immediate or very close contact with the surface is thus needed (see more details below). Consequently, only 'sessile' micro-organisms contained in the 'biofilm' are involved in the corrosion reaction. Such a biofilm is a very complex layer of a variety of active cells, macro-molecules of lipids, proteins, and polysaccharides [3].
- *Sulphide-generating:* sulphide generation is a general characteristic and a usual flag indicating tendency to MIC, as indicated by the frequent presence of iron sulphide and the smell of sulphide from a corroded area. Even if sulphate-reducing bacteria (SRB) are usually indicated as the corrosive bacteria, thio-sulphate-reducing bacteria (TRB) have also been shown to induce very severe localised corrosion subject to meeting favourable growth conditions [4–7]. Other sulfidogenic micro-organisms are also pointed out in the literature, in particular for thermophilic Archea. Other sources of MIC by acid-producing bacteria, nitrate-reducing bacteria, etc. are also well known, although not identified as leading causes in anaerobic oil and gas environments.
- *MIC-prone local environment:* the overall local environment accumulated between the biofilm and the steel surface, which makes the corrosion reaction more or less active: it is not only a matter of H_2S and sulphides but also the supply of corrosive species (CO_2, H_2S, acidity, etc.) and other constituents either promoting localised corrosion or preventing it. The diversity of the microbial environment and the synergies between various types of micro-organisms contribute to this local environment.

- *Appropriate environmental conditions:* this involves the environmental contributors outside of the biofilm itself, such as the temperature, the flow velocity, the concentration in sulphate and nutrients, the anaerobic nature of the environment and of course the applied mitigation solutions. It is well known that the presence of supposedly corrosive micro-organisms does not necessarily mean that MIC is taking place: whether they are active or not is definitely related to such environmental conditions. It is probably these conditions which are the most decisive in explaining whether MIC occurs in a given condition.
- *Abiotic corrosivity (i.e. without any microbial activity):* not all the corrosive capacity is necessarily provided by the micro-organisms. One of the reasons why MIC can be so severe is that MIC usually benefits from the existing abiotic corrosivity due to the acidic CO_2-loaded produced water. When dealing with MIC, it is thus also necessary to deal with this corrosivity, as already highlighted in the conclusion of the previous paragraph.

11.2.1 Summary of Key Parameters

Having considered all parameters outlined in Figure 11.1 [8], and 11.4 therefore, it can be said that:

- MIC is due to the biofilm formed on the metal surface.
- Internal MIC in hydrocarbon production facilities is essentially related to sulphides generated by micro-organisms.
- SRB are not the sole contributors that can be involved: to talk of bio-corrosion by SRB is an over-simplification.
- Whether the appropriate environmental conditions for microbial activity are present is probably the most important MIC factor.
- MIC is not separated from other corrosion mechanisms: it is not because humans like a unique and strictly defined cause for their issues that nature complies!

11.3 The Motive for Promotion of Corrosion by Micro-organisms

In order to describe the MIC phenomena in a simple and easily understandable way, it is helpful to address the subject systematically and independently. First, a description of what drives corrosion at the steel surface and interface is provided. This is followed by a description of what drives microbial activity inside the biofilms. Finally, what may constitute a bridge between the two is discussed, without entering into any complex terminology and modelling from any of the two microbial and corrosion communities. The approach here is intended to be straightforward, as it only considers the two ends of the microbial corrosion involving sulphate reduction (or any comparable species), while completely discarding all intermediate steps: at one end, iron is oxidised and, at the other end, sulphate is reduced. Therefore, the reason behind the promotion of corrosion can be described in the following sections.

11.3.1 The Corrosion Process

It is acknowledged that MIC is no different from other corrosion mechanisms experienced in wet environments: corrosion is basically an oxidation and dissolution of iron to the environment (Eq. 11.1):

$$(Fe \rightarrow Fe^{++} + 2e^{-}) \tag{11.1}$$

- It is also fully agreed that this transfer of Fe^{++} to the solution must be complemented by a coupled cathodic reaction 'picking up' the two free electrons associated with the above oxidation reaction (reduction).
- Basically, from the metal/ fluid interface, MIC is primarily a matter of charge transfer, combined with mass transfer limitations when the supply of reacting species is slower than their oxidation or reduction kinetics.
- The amount and the kinetic supply of corrosive species beneath the biofilm can thus be a corrosion driver, as long as a 'uniform corrosion mechanism' is considered (i.e. with similar and opposite anodic and cathodic reactions at the same location).
- A coupling between a corroding surface and the surrounding surfaces can also take place, leading to higher local corrosion rates by a kind of 'localised corrosion mechanism', sustained by cathodic reactions also taking place on the surrounding surfaces: nothing new or surprising in corrosion!
- Last but not least, the corrosion process may also be controlled by the corrosion layer built in the corroding area.

11.3.2 Microbial Activity Inside the Biofilms

- As a simple summary of their complex activities, micro-organisms also reduce species from their environment (e.g. sulphates to sulphides for SRB) and oxidise other species (e.g. organic carbon to CO_2), similar to human respiration and food consumption.
- These are the usually labelled 'electron acceptor' and 'electron donor' reactions recurrently pointed out in the MIC literature. As mentioned above, reaction kinetics may be limited by either the overall charge transfers involved in these reactions, or by the limited supply of reactive species.
- There are many potentially limiting factors of this microbial activity, particularly among the 'appropriate environmental conditions' listed above and the diversity of microbial communities present inside the biofilm. As far as sulphide generating micro-organisms are concerned, the sulphate content in the fluid may also be a mass transfer limitation of sulphate-reducing kinetics.
- As long as it is assumed that sulphide generation is the leading corrosion driver, sulphate reduction kinetics are certainly decisive of the corrosion growth. Gu highlights the biocatalytic sulphate reduction induced by biofilms, which may enhance charge transfer-controlled reactions [9]. On the other hand, in a situation of limited mass transfer, the diffusion of sulphate or organic carbon to the biofilm will drive the corrosion rate [9].
- It has also been shown that SRB were not only able to provide H_2S to the local corrosive media below the biofilm, but might also act as a pH regulator [6].

- Bonifay et al. [10] also highlight that, among the large variety of metabolites produced by micro-organisms, some may be decisive of the extent of corrosion occurring from one to another piece of equipment. This difference may be either caused by the supply of additional corrosive behaviour or by preventing the build-up of a protective corrosion layer.
- Finally, this proposed simple MIC mechanism considers that the key contribution of the biofilm to MIC is to provide the steel surface with the above-mentioned 'MIC prone local environment': if it does, MIC will be severe and, in contrast, if it does not, MIC will not develop. The next step is about how the steel surface accommodates itself with what the biofilm provides.

11.3.3 Bridging Surface to Biofilm

As outlined above, the sole aim here is to propose five simple and non-exclusive mechanisms bridging the two parts discussed earlier and summarised in Figure 11.5. The respective mechanisms and steps are outlined herewith:

1) *Bridging mechanisms 1–3* correspond to the ultimate chemical and electrochemical exchanges exclusively occurring inside the local corroding cell where the biofilm is active, through a direct chemical and electrochemical exchange between the biofilm, the local intermediate environment, and the steel surface. In other words, a 'general corrosion' on a local surface, thus leading to a hemispherical profile (all directions at the same rate).

   - *Mechanism 1 is the simplest,* as it assumes a completely de-coupled interaction between the biofilm and the steel surface. In this, microbial activity provides

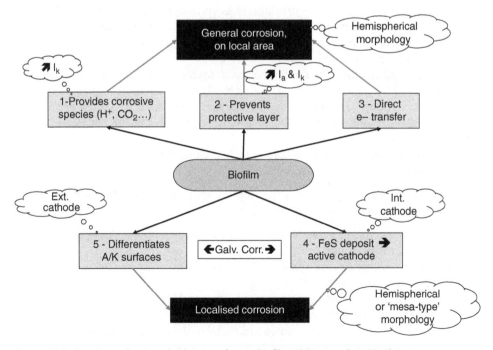

Figure 11.5 Simple mechanisms bridging surface to biofilm. (A/K = anode/cathode).

corrosive species at the steel surface (acidity, CO_2, etc.), i.e. a cathodic current, while the anodic reaction/corrosion consumes these corrosive species. This is sometimes called chemical MIC (CMIC) [11, 12].

- *Mechanism 2* considers the possibility that the first role of what is produced by the biofilm is to prevent the formation of a protective layer on the steel surface (e.g. via particular metabolomes, as indicated earlier), thus allowing the abiotic corrosivity to corrode with its full capacity, without being reduced by a protective corrosion layer as on the rest of the surface. In practice, this mechanism can be complementary to the previous one.
- *Mechanism 3* refers to the so-called 'electro-active biofilms' and to the said CMIC [11, 13]. This tentative mechanism considers a direct electron transfer (e-transfer) via nano-wire connections between bacteria and the metal surface [11, 14, 15]. The occurrence of such electrical transfer requires that the electrical connection to the metal is maintained over time, despite the likely build-up of a corrosion product inside the corroding cell: the author has no documented evidence that this can be a stable and significant process over time. As opposed to mechanism 1, what occurs in the biofilm is coupled with what occurs on the steel surface. This mechanism can also be complementary to this first mechanism.

2) *Bridging mechanisms 4 and 5* (Figure 11.5) consider a galvanic coupling of the corroding surface, leading to a differentiation between the corroding cell and a larger cathodic surface, i.e. with an enhanced corrosion rate.

- *Mechanism 4* considers a local galvanic coupling as described herewith:

  - *Microbial activity inside biofilm* → provides corrosive species and sulphides below the biofilm.
  - *Corrosion provides Fe^{++}* → precipitates as iron sulphide from sulphides delivered by micro-organisms inside and on top of the local corroding cell.
  - *Iron sulphide acts as an active cathode*, producing a local galvanic coupling with the corroding surface.
  - This mechanism is certainly the most frequently proposed to explain the localised MIC morphology. With this mechanism, a significant part of the galvanic coupling is internal to the corroding cell itself, apart from the iron sulphide possibly present on top of the corroding cell which may be coupled with the bulk water.
  - This mechanism is compatible with a hemispherical morphology, i.e. a similar local growth rate in all directions.

- *Mechanism 5* envisages a galvanic coupling with the surrounding surface, out of the corroding cell as described herewith:

  - *Microbial activity inside the biofilm* → provides specific species at the steel surface (corrosive or not).
  - *Which promote differentiation with the rest of the surface* (in pH, sulphide content, corrosion layer, etc.).
  - *Leading to a localised corrosion* between the local biofilm influenced surface and the surrounding surface.
  - In this mechanism, which can be additional to the previous one, local corrosion is enhanced by taking a 'benefit' from the external abiotic corrosion mentioned in the previous paragraph.

– Depending on the extent of the galvanic effect and the coupling distance, the morphology can either be hemispherical (long-distance coupling) or rather of a 'mesa-type' profile (short-distance coupling close to the edge of the corroding surface).

11.3.4 Summary Mechanism

In practice, it is likely that several of these 'bridging mechanisms' are jointly involved. As far as the 'electroactive biofilm' is concerned, there is little or no evidence that this effect is of noticeable contribution in real corroding situations.

It is also worth highlighting that, even if the simplifying objective in this chapter tended to separate the biofilm from the surface by the intermediate local corrosive fluid, the reality is certainly that the biofilm is part of the metal–fluid interface, as highlighted by Beech and Sunner [16]. As such, this biofilm certainly modifies to some extent the interfacial properties, hence the electrochemical kinetics which may take place between the corrosive fluid and the metal. On the other hand, it is regularly observed that, under MIC-prone conditions, local corrosion rates below the biofilms can be at least as high as and even higher than the abiotic corrosivity of the bulk fluid: this means that the interfacial properties have not really been so dramatically affected, at least from an inhibiting side.

In brief, it can be concluded that:

- As long as MIC is limited to a local effect inside the 'corroding cell' induced by a local intense microbial activity, the worst corrosion rate is directly related to the microbial activity inside the biofilm, i.e. finally with the sulphate-reducing rate inside this cell (under either mass or charge transfer limitation inside the biofilm). This limitation applies whatever the charge transfer mechanism to the steel surface is via intermediate chemicals or electronics.
- If any galvanic coupling occurs between the corroding area and any surrounding surface (iron sulphide, non-bioactive surrounding surface), the MIC rate can also be driven by the abiotic corrosivity of the bulk fluid, i.e. by the CO_2 or $H_2S + CO_2$ corrosivity.
- From an operating company or end-user point of view, it is strongly suspected that an enhancement by galvanic effect is frequently experienced. This is discussed in Section 11.4.

11.4 Most Susceptible Locations and Conditions

It has been outlined above that the environmental conditions wherein MIC can occur are probably mostly controlled irrespective of its occurrence in a given operating conditions. There are too many factors influencing this, including but not limited to the microbial activity, such as the temperature, the pH, the salinity, the redox potential, the availability of major and minor nutrients, the flow velocity, the microbial diversity, and the corrosivity of the fluid. It thus looks somewhat illusory to try predicting from scratch whether a given environment can promote MIC. It is reasonable to conclude that all attempts to try predicting the ability of any environment to favour or prevent

Table 11.1 The likelihood of MIC threat occurrence: a brief overview.

MIC occurrence	Locations	Operating scenarios
Most likely environments and facilities	Oil-water networks Water injection networks Oil or produced water vessels and tanks Produced water piping Wet oil piping and low spots on closed drain networks Cooling water network	Operating temperature range within 20–50 °C Low fluid flow velocities Solid accumulation sites
Least likely environments and facilities	Gas production Gas compression processes Producing wells MEG networks	Highly alkaline produced water
Uncertain limits and conditions of occurrence	Temperature limits Flow velocity limits Sulphate limits Salinity limits Nutrients limits Effect of the abiotic corrosivity/ protectivity	0–15 °C temperature range Above 80 °C Liquid flow velocities in excess of $2\,m\,s^{-1}$ Sulphate in excess of $5\,mg\,l^{-1}$ No salt limit

MIC have failed, except for very restricted cases (e.g. for the temperature effect in a specific medium).

From the hydrocarbon producers' point of view, i.e. from companies regularly facing potential MIC threats, a more modest attempt can be to make a distinction listing typical cases and conditions under which MIC has been either regularly or rarely experienced. This is attempted in the list presented below, while by no means claiming to represent the exhaustive experience of operating companies. These are categorised into three groups, summarised in Table 11.1 and outlined in Sections 11.4.1–11.4.3.

11.4.1 Most MIC-Prone Environments and Facilities

- *Oil–water networks:* it is generally recognised that oil–water producing pipelines and process piping are the most susceptible facilities.
- *Water injection:* water injection networks also are among the usual MIC-affected candidates, particularly when a mixture of produced water and sea water is injected.
- *Favourable temperature range:* oil–water production and water injection facilities operating in the 20–50 °C range are among the most affected by MIC.
- *Favourable flow velocities:* low flow velocities, quasi-stagnant conditions and facilities with solid accumulation are particularly MIC-prone.
- *Vessels and tanks:* oil or produced water vessels, and particularly storage tanks, are potentially very prone to MIC because they frequently combine all the detrimental factors listed above. In practice, MIC issues are rarely experienced but this is only

because they are frequently protected by an internal organic coating and CP, i.e. subjected to a robust mitigation.

- *Piping:* produced water piping, dead ends of wet oil piping and low spots on closed drain networks are definitely among the weakest components of an oil–water treating facility.
- *Cooling water networks:* finally cooling water networks, either opened or closed, are also MIC-prone facilities.

11.4.2   Least MIC-Prone Environments and Facilities

- *Gas production:* previous operational experience demonstrates that gas-producing facilities are very rarely subjected to MIC though no preventive treatment is applied. Exceptions include a few mature gas pipelines with low flow velocities where reservoir water is being produced and in produced water piping and closed drain piping on gas treatment facilities. Among the environmental conditions preventing MIC there, it is assumed that high flow velocities and low salinities/low sulphate contents of condensed waters are the most decisive ones.
- *Gas compression processes:* as far as wet gas facilities are concerned, there has been no experience of MIC issues in compression processes: the fact that there is almost uniquely condensed water is certainly favourable, as well as the severe local heating of the wet gas through compressors.
- *Producing wells (oil and gas):* very little MIC experience is also noted in producing wells, even for oil–water producing wells, despite the fact that no preventive treatment is implemented. It is expected that this is due to a combination of detrimental flow regime (slug or annular flow) and also from higher average temperatures inside producing wells than on downstream pipelines and production facilities.
- *Mono-ethylene glycol (MEG) networks:* not surprisingly, no MIC issue has been experienced yet on lean or rich MEG networks.

11.4.3   Uncertain Limits and Conditions of Occurrence

- *Temperature limits:* even if it is frequently experienced that serious MIC issues occur in the 20–50 °C temperature range, it is more difficult to predict how serious MIC can be at lower and higher operation temperature conditions:
  - At the lower operating temperature conditions within the 0–15 °C range, very little MIC issues have been experienced to date, although its occurrence cannot be ruled out, albeit at lower rates: in other words, it is a matter of insufficient experience or of low risk.
  - At higher operating temperature conditions, the question arises whether thermophilic micro-organisms (above 50–60 °C) are still significantly corrosive. A conservative position would be to assume that MIC remains possible above 50 °C, although it is not clear up to what temperature. As an example, Cochrane [15] reports on the isolation of thermophilic bacteria showing an active behaviour up to 70 °C at least.
  - An 80 °C value is presently considered a risk limit by the author: while not nil, the risk is certainly much lower above such a temperature.

- *Flow velocity limits:* it is generally accepted that low flow velocities are favourable to MIC and hence the natural tendency of any engineer will be to ask for a 'critical liquid or water velocity' above which MIC is no longer an issue. Typical values in the range up to $2\,m\,s^{-1}$ are frequently referred to. However, this is only a realistic order of magnitude but by no means a strict guarantee of no MIC at higher fluid flow velocities. In particular, the critical velocity for biofilm development is certainly variable depending on the types of biofilms (the thicker being a priori the most sensitive). Such velocity tends to prevent a complete water wetting in an oil–water mixture with a low basic sediment and water (BSW < 5–10%) and also to prevent the accumulation of solid particles, if any. Therefore, this may provide conditions to minimise MIC risks. The $2\,m\,s^{-1}$ limit should thus be considered a risk limit, in a similar way to that described above for the temperature limit and not necessarily zero risk of MIC above $2\,m\,s^{-1}$, but significantly lower.
- *Sulphate limits:* similar to that for temperature and velocity limits, a limit in sulphate content is also desirable by designers, below which no MIC is potentially foreseen:
  - As long as MIC can be limited by the mass transfer, there is certainly a limit. However, as on one hand, mass transfer is also dependent on other factors and, on the other, a part of MIC damage is by galvanic coupling, it is unrealistic to give a strict guarantee on a fixed value.
  - Generally, a limit of $5\,mg\,l^{-1}$ is considered the limit between high and low risk areas.

- *Salinity limits:* MIC problems have been found on a pipeline transporting a saturated NaCl brine (>$300\,g\,l^{-1}$): no salt limit is thus considered for MIC.
- *Nutrients limits:* Whether low or high amounts of nutrients promote or minimise MIC is indeed a matter of scientific discussion. However, from an operational point of view, the environments totally free of nutrients are certainly very rare as soon as some oil has been in contact with a produced water. Consequently, except in a few cases or very controlled waters (e.g. potable water, desulfated sea water, cooling waters), it is prudent never to consider that MIC might be controlled because of a lack of nutrients.
- *Effect of the abiotic corrosivity/protectivity:* As long as MIC-prone environmental conditions are present, there are numerous cases of practical experience where MIC rates have exceeded by a factor of 5–10 the calculated abiotic corrosivity of the produced water (estimated with conventional CO_2 corrosion prediction tools, assuming no protective layer). A low abiotic corrosion rate does not guarantee a low MIC rate. On the other hand, a high abiotic corrosion rate is no indicator either: a high abiotic corrosivity tends to guarantee the worst corrosion outcome if MIC is also involved. On the other hand, it is also the author's experience that highly alkaline produced waters (>$2000–3000\,mg\,l^{-1}$ bicarbonate), which are known to show a very low CO_2 corrosion rate because of the easy protection of an iron carbonate layer, also show minor MIC rates despite serious bacterial contamination. The protection offered by this high bicarbonate content thus looks strong enough not to be locally compromised by the microbial activity.

11.4.4 Brief Overview

In conclusion, the above conditions can be summarised in Table 11.1 with notable points as follows:

- Few typical experiences are given where MIC is or is not a potential threat. Typical risk limits are also outlined for few operating parameters.
- There would be a definite interest and benefit in a joint effort from operators to sharing a list of operating conditions under which MIC has been faced. This would permit better appreciation of limits and a reiteration of what has been outlined earlier.

11.4.5 The Anticipated Damage Rate

It has been the usual rule of thumb that MIC on oil and gas facilities is in the $1-2\,mm\,y^{-1}$ range. However, failure cases at more than $10\,mm\,y^{-1}$ have also been experienced (cf. Figures 11.1–11.3). This was indeed in a situation where bacterial contamination was very high and where TRB were also present and active. Nevertheless, this case is definitely not unique and MIC rates can easily exceed this $1-2\,mm\,y^{-1}$ limit. In conclusion, as long as the most appropriate environmental conditions are met for an active microbial activity (see above), MIC rates up to $\sim 10\,mm\,y^{-1}$ can be expected.

11.5 Potential Prevention Measures

Depending on the type of equipment and the environmental conditions, MIC mitigation can either be:

- conditional, i.e. linked to a decisive indication (generally from microbial or corrosion monitoring), or
- systematic, i.e. not dependent on any prior indication.

Figure 11.6 summarises a list of usual mitigation measures, the typical equipment to which they apply to in oil and gas production/injection facilities and whether they are conditional or systematic. These are described further in Sections 11.5.1–11.5.5.

11.5.1 Biocide Treatments

Biocide treatments aim at controlling the activity of micro-organisms contained inside adherent biofilms to a sufficiently low level. This is the main reason why biocide treatments have essentially shifted to discontinuous periodic treatments by 'batch' treatments, at a high dosage (typically 300–1000 ppm vol. depending on the concentration of active component in the biocide, for four to six hours every one to two weeks). It has indeed been progressively acknowledged that such a high dosage treatment was more efficient and more cost-effective than continuous treatments at low dosage against micro-organisms protected by the biofilm 'umbrella'. It is also worth noting that a 500 ppm/5 hours/2 weeks discontinuous treatment is equivalent, from a biocide consumption point of view, to a continuous treatment at ~ 7 ppm vol: thus, less chemicals' consumption than continuous treatments at concentrations of 2–30 ppm vol.

Though this mitigation solution has definitely proved its efficacy, particularly in combination with periodic scrapping, its potential environmental impact is now an increasing challenge when disposal of the treated waters is required. Responses to the increasingly stringent regulations have progressively been developed by the industry. Among them, the most decisive have certainly been:

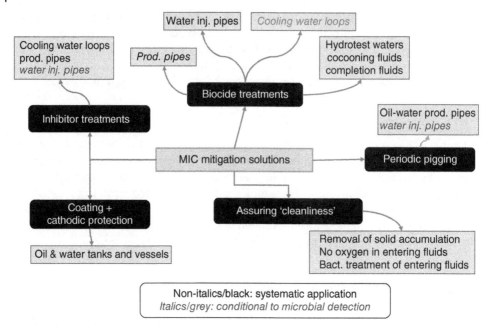

Figure 11.6 MIC mitigation measures/solutions.

- *Sharply minimising the disposed amounts of produced waters,* by a quasi-systematic reinjection of produced waters to their native reservoirs.
- *Using less environmentally impacting products:* the formal eco-toxicological quotation of all chemicals, per the OSPAR Convention [17] and related application documents, has permitted classifying chemicals used versus their environmental impact and specifying all products used accordingly.
- *Quantifying the environmental impact in case of disposal,* using dispersion models and considering not only the eco-toxicological properties of the disposed products but also their quantities and concentrations.
- *Measuring the environmental impact* in the area subjected to a potential contact with the disposed species.

Minimising the amounts of biocides used is also an action line, as long as this can be done without impairing the efficiency of the MIC mitigation, i.e. not increase the corrosion risk – these include:

- *Correct dosage:* the most promising, though not the easiest solution, would obviously be to use 'only the required amount/dosage, and only when needed', i.e. to fully optimise the injected dosage, the injection duration, and the periodicity of batch treatments. The essential condition to reach this objective is to have access to a reliable and quick monitoring of how MIC develops at all times when a treatment is applied: this requires answering several questions including:

  – Is it high or low before a treatment?
  – How much does it decrease after a treatment?
  – When does it re-increase to such an extent that a new batch is required?

Unfortunately, the oil and gas and the corrosion monitoring industries are still far from having access to such monitoring capacities, as discussed in Section 11.6.

- *The use of 'boosters':* this is a recent area of research which may enhance the efficiency of more conventional biocides: recent publications have highlighted the boosting capacities of D-amino acids (DAA) [18, 19], of essential oils [20] in combination with other biocides. Tentatively such boosters might allow the dosage of conventional biocides (THPS, *gluteraldehydes*, etc.) to be reduced by a factor of two to four, i.e. with an equivalent reduction in the use of conventional biocides. Even if this approach is not yet at an advanced stage of development and application, it is expected to be a growing area of activity if positive responses start being obtained on production sites.

11.5.2 Periodic Pigging

For pipelines, sending a mechanical pig (with hard plastic plates) to clean wax/debris inside the pipe prior to every biocide injection is considered an essential measure. This treatment flushes away part of the biofilm and deposited solids, if any, hence making quicker and easier the action of biocides on the surface. In practice, it is hard to tell what, between the biocide treatment and the scraping, is the most prevalent mitigating effect: it is not ruled out that the scraping alone does more than half the job in many cases, though it is not considered a sufficient solution for any serious MIC risk. On the other hand, what is widely acknowledged is that not performing such scraping when it is possible, even at the price of some operational constraints, impairs the likelihood of controlling MIC in a satisfactory way.

11.5.3 Inhibitor Treatments

When facing any serious MIC issue, assuring a rigorous corrosion inhibitor treatment (i.e. permanent and at an appropriate dosage) is certainly a prudent attitude. As already outlined, MIC is not so completely separate from other abiotic corrosion mechanisms to be solely treated by biocides and scraping. It is solely in situations of very moderate abiotic corrosivity (e.g. deaerated and degassed water) that such corrosion inhibitor treatment would be considered over-conservative.

11.5.4 Cleanliness

What is included in 'cleanliness' is a hygienic operational attitude which consists of taking care not to let solids, oxygen, waste, dirty waters, etc. enter the production or injection facilities. It also consists of regularly removing solid deposits, and cleaning and drying vessels before closing them. This 'clean' attitude can be a way to prevent or at least minimise serious internal contamination which is then hard to remove. It is thus classified one of the MIC mitigation solutions. The periodic scraping discussed earlier is definitely one element of this 'clean' attitude.

11.5.5 Cathodic Protection and Coatings

Combining cathodic protection and coating solution is one of the simplest, very cost-effective, and most efficient MIC mitigation methods for vessels and tanks [21]. There are still numerous discussions about the necessary protection potential to be achieved to assure a complete MIC mitigation (−950 mV/SCE being currently indicated in most standards [22]). It is also an efficient solution to protect buried equipment from external MIC. In practice, its main limitations are:

- shielding effects from some coatings, which prevent the required potential to apply where needed below damages coatings, where microbial activity still works;
- obviously the difficulty of applying any cathodic protection inside tubular equipment (piping or pipelines) where there is no continuous water phase to drive the protection current, such as oil−water or gas-oil- water mixtures.

11.6   Means of Monitoring

When it comes to MIC monitoring in operations, it is conceivable to conclude two extreme and opposing summaries, both true to some extent:

1) *There is formally no specific corrosion monitoring technique for MIC – if it is only about monitoring the corrosion itself* – except for inspection, which is hardly a monitoring solution, there is no real MIC-specific corrosion monitoring solution in widespread development. Few solutions providing a specific electrochemical response to MIC have been proposed [23–25]. However, even without considering their real performance as corrosion monitoring devices, the usual difficulty in performing reliable electrochemical measurements in the presence of oil, even at trace amounts, poses the recurrent question of their limited application domain in the oil and gas production.
2) *Dealing with MIC monitoring techniques is not a matter of one paragraph but rather of books* – as long as some details are required, particularly when it comes to the diversity and complexity of the various molecular microbiological methodologies now available.

Between these two extreme positions, the first question that an operator facing a MIC problem should have to answer about monitoring is certainly one of the two following options:

   Option 1: Keep things simple and rustic: This attitude essentially consists in working with selective microbial numeration by various serial dilution solutions (according to NACE TM0194 standard [26]), complemented with corrosion coupons and probes and on-line or off-line wall thickness measurements for the corrosion side, bearing in mind that these methods are not MIC-specific. Advantages and drawbacks of this option are summarised in Table 11.2.

   Option 2: a move to more modern, though more complex and more committed molecular microbiological techniques, dealing with genomic characterisation: There are now a number of such solutions allowing, in a more or less quantifying way, either the variety of micro-organisms present or specific genes or functionalities of interest (e.g. the dsrAB gene involved in sulphate reduction) to be determined. The individual cost of most of these analyses is well known to have decreased by thousands, in less than 20 years, moving the use of these Molecular Microbiological Methods (MMM) solutions from fundamental academic studies to field monitoring applications. Positive uses of these techniques now are starting to be reported on oil and gas fields,

Table 11.2 MIC monitoring.

		Option	
No	Definition	Advantages	Drawbacks
1	Basic approach (keep things simple – (microbial numerations + corrosion monitoring)	• Simplicity, moderate cost, easy to be done on site, rather specific to selected types of micro-organisms (e.g. SRB, TRB, NRB, etc.) • The ability to consider sessile bacteria present on a surface or on solid samples	• 2–3 weeks reaction time: it is not feasible optimising any periodic treatment done every 1–2 weeks with such monitoring • It is also frequently argued that standard numeration kits show a very low sensitivity and may ignore 50–90% of the active species present. This position is not necessarily agreed, as long as few precautions are taken: using high sensitivity kits (e.g. Magot [27]), performing immediate on-site inoculation, preventing any aeration prior to inoculation.
2	Modern methods (molecular microbiology methods)	Much more detailed information made accessible by these techniques than with cultivation techniques (as long as the appropriate ones are selected), by more complete information of micro-organisms, genes, functionalities present. A very short response time – as soon as the sample arrives to the specialised laboratory (hours to day). New and quickly evolving techniques → Likely improvements in the ease of use, the quality of deliverables and the domains of application in a near future.	• No on-site measuring devices yet → Inappropriate for sites located far away from specialised laboratories equipped for such measurements. • High technical level required → Requires competent specialised resources for performance and evaluation. • Large amount of information provided, • No general and easy evaluation criteria yet → What to do with these results? • Not a better discrimination between 'corrosion active' and passive micro-organisms than the previous solution → No direct knowledge whether MIC is a true concern or not.

particularly in the North Sea sector. Advantages and drawbacks of this option are summarised in Table 11.2.

The question of whether or not to move to the potential complexity and added monetary cost and effectiveness of solutions is certainly first related to the proximity of fully equipped and specialised laboratories in close proximity to the concerned assets. This situation may probably develop in the foreseeable future, but the minimum conditions are that:

- miniaturised and simplified testing devices are made available for oil and gas applications, which can be used on-site or close to the site without necessitating an intense microbiological training for the operators;
- simple and consistent evaluation criteria are sorted out, not requiring microbiological experts to evaluate any single result.

Of more importance and still under extensive discussion is whether the knowledge of part or all of the microbial species and genes in place can be related to the corrosion which occurs. Knowledge transfer to all those involved is paramount here.

Another recently proposed solution is to look for specific metabolites related to the corrosion activity of the biofilm [10]. Though this solution might be a long-term one, preliminary promising results have shown a significant differentiation in the type of metabolites found in known corrosive and non-corrosive conditions, from chemical analyses instead of genomic analyses, as indicated earlier: in such an approach, 'who does the job' is still unclear but by-products showing that 'the job is done' can be measured.

Finally, as noted earlier, it is not one of the objectives of this chapter to provide more detail on all of these techniques. Dedicated chapters in a few recent books cited by Skovhus [12] and EFC Publications 22 and 66 on this matter constitute very helpful summaries of the various solutions, particularly for the various solutions included in the evolving molecular microbiology domain.

It should be emphasised that there is great potential for quick progress in the miniaturisation and simplification of measuring devices for genomic and metabolic analyses. It is likely that some of the very complex laboratory solutions used until now will become regular site-monitoring solutions. The main two constraints are, on the one hand, that simple and accurate evaluation criteria are defined for such measurements, and, on the other, that the business is sufficient to justify the development and production of such tools at a reasonable price.

11.7 Summary

MIC remains a major corrosion threat in hydrocarbon production. The subject has been scrutinised methodically from the operating companies' viewpoint where the threat can pose significant challenges. The focus and aim here have been to address the subject by describing its features and locations where it is most likely to occur, focused on engineering aspects of what, why, when, where, and how to mitigate. Key conditions necessary for the occurrence of internal MIC of CLAS equipment in most hydrocarbon production facilities are summarised with a description of each parameter and its significance.

A brief distinction has been made listing typical cases and conditions under which MIC has been regularly experienced against cases when it is rarely experienced or where there are uncertainties about its occurrence. An attempt has been made to characterise these locations in terms of operating conditions and scenarios. In addition, methods of mitigation and means of monitoring are dealt with in brief. The chapter is by no means exhaustive and the battle remains ongoing to address the subject in a more explicit manner.

References

1 ISO, Corrosion of metals and alloys: basic terms and definitions. ISO 8044, 2015.
2 S Shapira, Microbiologically influenced corrosion (MIC) in offshore systems: common problems and misunderstandings. Presentation given at the INSSOK Corrosion Seminar, Kuala Lumpur, 26–27 January 2010.
3 Beech, Y., Sztler, M., Smith, W.L. et al. (2014). *Biofilm and Biocorrosion*, EFC Publication No. 66. Elsevier.
4 Campaignolle, X., Festy, D., and Crolet, J.L. (1997). *A Search for the Risk Factors Involved in the Carbon Steel Corrosion Induced by Sulfidogenic Bacteria*. The Institute of Materials.
5 Crolet, J.L. (1992). From biology and corrosion to biocorrosion. *Oceanological Acta* 15 (1): 87–94.
6 J L Crolet, S Daumas, and M Magot, pH regulation by sulphate reducing bacteria. NACE Corrosion Conference, Paper No. 303, 1993.
7 Crolet, J.L. and Magot, M. (2006). Observation of non-SRB sulfidogenic bacteria from oilfield production facilities. *Materials Performance* 35 (3): 60–64.
8 Total, In-house MIC photos from oil–water pipelines operated at that time, 1989.
9 T Gu and D Xu, Demystifying MIC mechanisms. NACE Corrosion Conference, Paper No 10213, 2010.
10 Bonifay, V., Wawrik, B., Sunner, J. et al. (2017). Metabolomic and metagenomic analysis of two crude oil production pipelines experiencing differential rates of corrosion. *Frontiers in Microbiology* 8 (140).
11 Enning, D. and Venzlaff, H. (2012). Marine sulphate-reducing bacteria cause serious corrosion of iron under electroconductive biogenic mineral crust. *Environmental Microbiology* 14 (7): 1772–1787.
12 Skovhus, T.L., Lee, J.S., and Little, B.J. (2017). Predominant MIC mechanisms in the oil and gas industry. In: *MIC in the Upstream Oil and Gas Industry* (ed. T.S. Skovhus, D. Enning and J.S. Lee), 75–86. CRC Press.
13 Basséguy, R., Délia, M.L., Erabel, B., and Bergel, A. (2014). Electroactive biofilms. In: *Understanding Biocorrosion: Fundamentals and Application* (ed. T. Liengen, D. Féron, R. Basséguy and I.B. Beech), 107–133. Elsevier.
14 Dinh, H.T., Kuever, J., Mussmann, M. et al. (2004). Iron corrosion by novel anaerobic microorganisms. *Nature* 427: 829–832.
15 W J Cochrane and P S Jones, Studies on the thermophilic sulfate-reducing bacteria from a souring North Sea oil field. SPE European Petroleum conference, SPE Paper No. 18368, London, 1988.
16 Beech, I.B. and Sunner, J. (2004). Biocorrosion: towards understanding interactions between biofilms and metals. *Current Opinion in Biotechnology* 15: 181–186.
17 OSPAR Convention, Convention for the Protection of the Marine Environment of the North Sea Atlantic, (1992 + further amendments) 1992.
18 Xu, D., Li, Y., and Gu, T. (2014). D-methionine as a biofilm dispersal signaling molecule enhanced tetrakis hydroxymethyl phosphonium sulfate mitigation of Desulfovibrio vulgaris biofilm and biocorrosion pitting. *Materials and Corrosion* 65 (8): 837–845.

19 Y. Li, R. Jia, HH. Al-Mahamedt, et al., Enhanced biofilm mitigation of field biofilm consortia by a mixture of D-amino acids. *Frontiers in Microbiology*, 7, Article ID 896, 1–13, 2016.

20 T. Lisowski and K. Esser, Synergistic disinfecting compositions with essential oils. Patent n° WO 2012/101129 A1, 2012.

21 Wilson, S.L. and Jack, T.O. (2017). MIC mitigation: coatings and cathodic protection. In: *MIC in the Upstream Oil and Gas Industry* (ed. T.S. Skovhus, D. Enning and J.S. Lee), 255–276. CRC Press.

22 NACE, Control of external corrosion on underground or submerged metallic piping systems, NACE Standard SP0169, 2013.

23 Cotiche, V. (2012). Biocorrosion probe monitors MIC of C steel. *Materials Performance* 51 (5): 14–16.

24 GJ Licina and G Nekoksa, The influence of water chemistry and biocide additions on the response of an on-line biofilm monitor. NACE Corrosion Conference, Paper No. 527, 1995.

25 Pavanello, G., Faimali, M., Pittore, M. et al. (2011). Exploiting a new electrochemical sensor for biofilm monitoring and water treatment optimization. *Water Research* 45: 1651–1658.

26 NACE, Field monitoring of bacterial growth in oil and gas systems, NACE Standard TM0194, 2014.

27 M Magot and G Ravot, Method of detecting sulphate-reducing bacteria. US patent n° 6531281 B1, 2003.

Bibliography

Crolet, J.L. (2005). Microbial corrosion in the oil industry: a corrosionist's view. In: *Petroleum Microbiology* (ed. B. Ollivier and M. Magot), 143–170. Washington, DC: ASM Press.

Liengren, T., Féron, D., Basséguy, R., and Beech, I. (eds.) (2014). *Understanding Biocorrosion: Fundamentals and Applications*, EFC Publication No 66. Woodhead Publishing.

Ollivier, B. and Magot, M. (2005). *Petroleum Microbiology*. ASM Press.

NACE, The role of bacteria in the corrosion of oil field equipment. TPC Publication No. 3, 1976.

Thierry, D. (ed.) (1997). *Aspects of Microbially Induced Corrosion*, European Federation of Corrosion Publication No. 22. Maney Publishing.

Figure 9.2 Completely disbonded three-layer PP coatings removed from a gas pipeline.

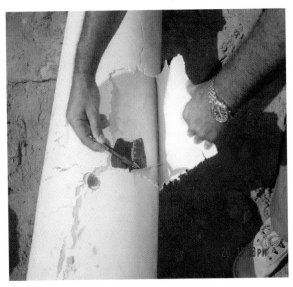

Figure 9.3 Cracking and disbondment of the three-layer PP coating due to thermo-oxidative degradation.

Corrosion and Materials in Hydrocarbon Production: A Compendium of Operational and Engineering Aspects, First Edition. Bijan Kermani and Don Harrop.
© 2019 John Wiley & Sons Ltd. This Work is a co-publication between John Wiley & Sons Ltd and ASME Press.

Figure 9.4 Failure of a flame-sprayed PP field joint coating on a three-layer PP-coated gas pipeline.

Figure 9.5 Application of hot PP tape to the weld joint of three-layer PP-coated pipeline.

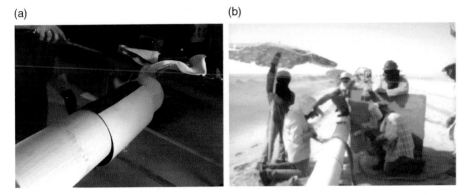

Figure 9.6 (a) Heat shrink sleeve coating on a weld joint in the desert; (b) application.

Figure 9.7 Field insertion of HDPE liner in a steel flowline.

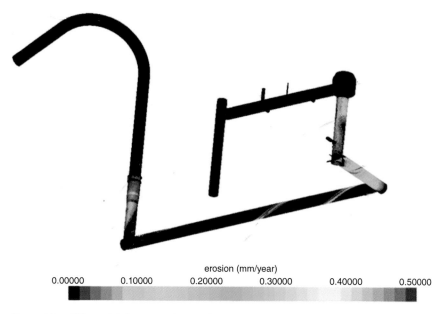

erosion (mm/year)

| 0.00000 | 0.10000 | 0.20000 | 0.30000 | 0.40000 | 0.50000 |

Figure 10.1 CFD model showing relative erosion rates in well-head piping. Red colour denotes the highest rate.

Figure 10.2 Access fitting: kit and example installation.

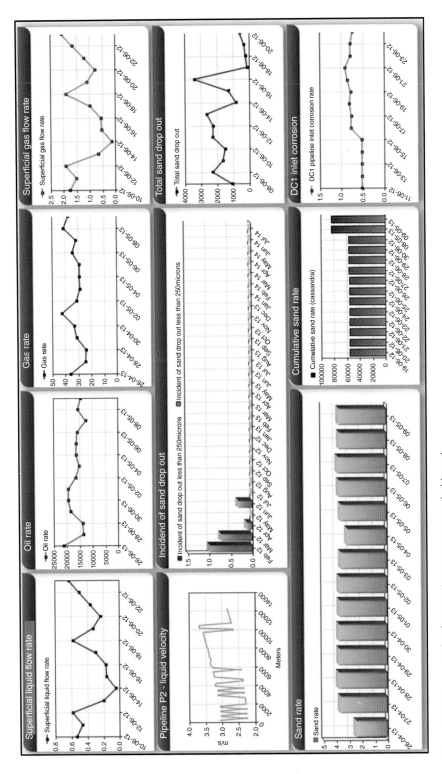

Figure 10.4 An example of a corrosion management dashboard.

Figure 10.5 A one-person-operated, portable RT unit in use.

(a)

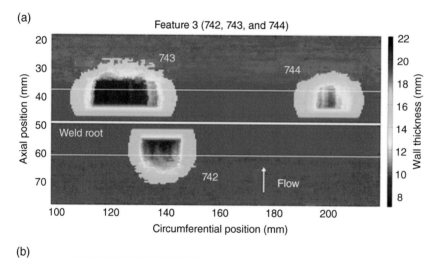

(b)

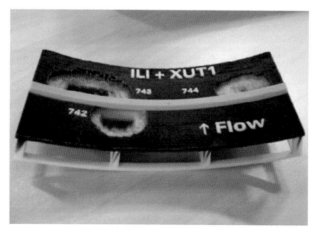

Figure 10.6 (a) Combined external and ILI UT data showing pipeline anomalies adjacent to a weld; and (b) a 3D-printed model of the pipeline anomalies.

Figure 11.1 Typical features of MIC type damage characteristic of localised morphologies.

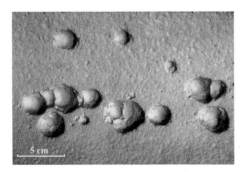

Figure 11.2 Localised hemispherical damage.

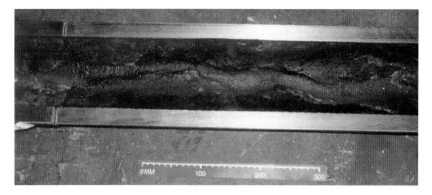

Figure 11.3 Groove corrosion promoted by microbial activity. *Source:* [2].

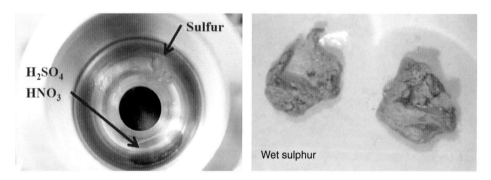

Figure 12.3 H_2SO_4, (7 M) HNO_3 (0.2 M) and elemental sulphur formed in corrosion experiment performed at 45 °C, 100 bar and CO_2 composition, as given in last column in Table 12.1, *Source:* [11].

(a)   (b)

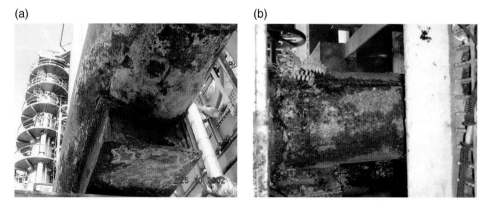

Figure 13.1 (a) and (b) typical examples of CUI showing devastating consequence of the corrosion threat.

Figure 13.6 A typical example of carbon steel sweating CUI damage accelerated by galvanic corrosion at a carbon steel/stainless steel flange material specification change.

Figure 13.7 Isolated CUI of piping with expanded pearlite insulation.

Figure 15.1 An example of heat ageing of a rubber seal.

(a) (b)

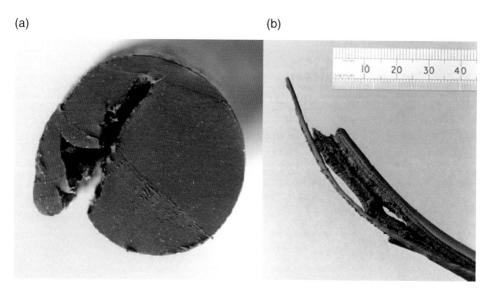

Figure 15.2 Examples of extrusion damage to rubber seals.

(a) (b)

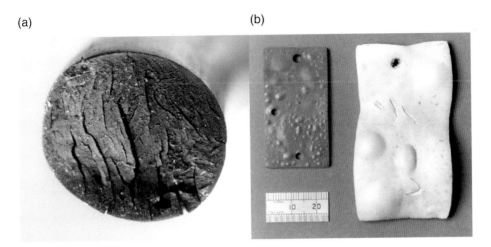

Figure 15.3 Examples of decompression damage to rubber.

Figure 15.4 An example of roller box reduction equipment.

1) Stainless steel carcass
2) Polymer fluid barrier
3) Carbon steel pressure armour
4) Anti wear/ birdcaging tapes
5) Carbon steel tensile armour
6) Polymer external sheath

Annulus

Figure 15.5 Schematic drawing of an unbonded flexible pipe. *Source:* Reproduced with kind permission from TechnipFMC, France.

Figure 17.2 An example of bottom of line groove corrosion.

Table 18.1 IM dashboard: sea water injection system

Monitoring/Mitigation		Week#12	Week#13	Week#14	Frequency/Target	Corrective action
Hypochlorite injection, Continuous, SW Lift Pump	Status PMvT				Continuous @ 20 ppm	
SRB/GAB (Planktonic) u/s Multimedia Filter	Status/PMvT		*N/A*		Bi-weekly/10 SRB per sample & 100 GAB ml^{-1}	
Chlorine, u/s Multimedia Filter	Status PMvT				Daily/0.5–1 ppm	
Fe, u/s Multimedia Filter	Status PMvT	*0.2 ppm*	*N/A*	*0.2 ppm*	Bi-weekly/Flat Trend	
TSS, u/s Multimedia Filter	Status PMvT	*10 mg l^{-1}*	*No sample*	*5 mg l^{-1}*	Daily/ <3 mg l^{-1}	
Cl$_2$/Bisulphite (OS) Continuous in VDAT sump	Status PMvT	*3 ppm*	*1 ppm*	*5 ppm*	4 ppm when SRU operating	
DO, Orbisphere d/s of VDTA	Status PMvT		*Offline*	*12 ppm*	Online/≤10 ppb	
Orbisphere calibration	Status		*Offline*	*???*	Every 30 days	
Biocide treatment, u/s SRU	Status/PMvT	200 ppm	200 ppm	400 ppm	1 hour/3 days @ 200–400 ppm	
Bisulphite (OS) residual, u/s SRU cartridge filters	Status/PMvT				Twice daily/0.64–2.56 mg l^{-1}	
Cl$_2$, u/s of SRP Cartridge Filters	Status/PMvT				Twice daily/*zero*	
DO, CHEMetrics (manual) SRU product header	Status/PMvT				Twice daily/≤10 ppb	
SRB/GAB (Planktonic) SRU product header	Status/PMvT		*N/A*	*No sample*	Bi-weekly/10 SRB/sample & 100 GAB ml^{-1}	
SRB/GAB (Sessile) SRU product header	Status/PMvT		*N/A*	*No sample*	Bi-weekly/10 SRB/sample & 103 GAB ml^{-1}	
ER corrosion probe, d/s water injection pump				0.2 mm year^{-1}	Online/Running average ≤0.1 mm year^{-1}	
Corrosion coupon, d/s water injection pump & Last coupon	Status/PMvT	Coupon Change due Week # 50 0.07 mm year^{-1}, no pitting			Yearly/ ≤0.1 mm year^{-1}, nom pitting	
DO, CHEMets (manual) d/s water injection pump	Status/PMvT				Twice daily/ ≤10 ppb	
Water velocity - in any given pipe section in system	PMvT				2 < Velocity < 10 m s^{-1}	

Green	Orange	Red	N/A

TSS = total suspended solids; PMvT = performance-measured versus target; VDAT = vacuum de-aeration tower; OS = oxygen scavenger; SRU = sulphate removal unit; DO = dissolved oxygen.

CHEMetrics is a company making a colorimetric DO test kit ampules charged with reactant that when filled with a sample of treated sea water colours according to the level of DO present. It is a quick visual means of measuring DO by comparing the colour of the tested water sample against a colour chart. (https://www.chemetrics.com/index.php?route=common/home). CHEMets are most commonly used in the upstream, being quick and simple to use albeit subject to a degree of user judgement on the colour change but a good means of correlation with the on-stream DO sensors. There may be other makers of such kits but the author is not aware of them.

Figure 18.3 A GIS map of a pipeline showing inspection data.

Figure 18.4 A 3D model of a processing facility created using 3D laser scanning.

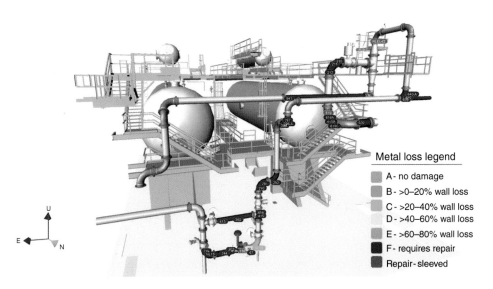

Metal loss legend

A - no damage
B - >0–20% wall loss
C - >20–40% wall loss
D - >40–60% wall loss
E - >60–80% wall loss
F - requires repair
Repair - sleeved

Figure 18.5 A 3D model of indoor facility piping showing inspection data.

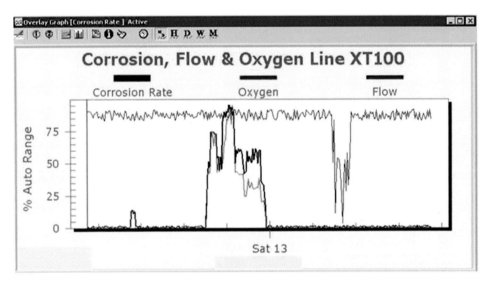

Figure 18.6 ER probe data from water injection line.

12

Dense Phase CO$_2$ Corrosion

With ever growing environmental constraints, global warming, and public awareness, there is an increasing incentive to reduce carbon emissions. The 'Blue Map Scenario' compiled and published by the International Energy Agency (IEA) for the abatement of climate change highlights a key need for the transportation and placement of CO$_2$ that is generated by associated industries, underground. Carbon capture and storage (CCS) is the means by which CO$_2$ is captured and compressed. This CO$_2$ then needs to be transported to a long-term storage site. In principle, transmission may be accomplished using pipelines, tankers, trains, trucks, compressed gas cylinders, as CO$_2$ hydrate, or as solid dry ice. However, only pipeline and tanker transmission are reasonable options for the large quantities of CO$_2$ associated with power stations, other industry activities, and hydrocarbon production. Effective transportation is justifiable through transmission of dense phase CO$_2$. This chapter combines aspects of materials and corrosion for the transportation of dense phase CO$_2$, outlining means of corrosion prediction, the materials, and the limits of application and any technology gaps that currently exist.

12.1 Background

Following the 'Blue Map Scenario' [1] for the abatement of climate change, it is estimated that about $10\,\text{Gtons}\,\text{year}^{-1}$ ($10^{12}\ \text{kg}\,\text{year}^{-1}$) of CO$_2$ needs to be safely transported and stored underground by 2050. The majority of the CO$_2$ will be transported by pipelines, as this is by far the most cost-effective and logistically robust option, although tanker transport is considered for smaller point sources. It is therefore, estimated that this requires the construction of about 3000 12-inch diameter (or 1000 20-inch diameter) pipelines, assuming a flow velocity of $1.5\,\text{m}\,\text{s}^{-1}$. Bearing in mind the extent of such investment, the only viable and cost-effective material that can be used for such an extensive pipeline network is carbon and low alloy steel (CLAS).

The typical operating window for the transportation of CO$_2$ using pipeline and tankers is indicated in the CO$_2$ phase diagram depicting the stability of the CO$_2$ at different pressure/temperature conditions (Figure 12.1). The viable option is to predominately

Corrosion and Materials in Hydrocarbon Production: A Compendium of Operational and Engineering Aspects, First Edition. Bijan Kermani and Don Harrop.
© 2019 John Wiley & Sons Ltd. This Work is a co-publication between John Wiley & Sons Ltd and ASME Press.

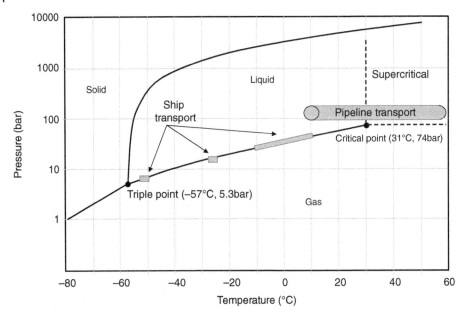

Figure 12.1 CO_2 phase diagram and typical pressures and temperatures for pipeline and ship transport.

transport CO_2 as dense phase (in the liquid or supercritical phase) and sufficiently dehydrated to avoid hydrate formation. Hydrate forms readily at about 11 °C in pure CO_2 when a free water phase is present.

Some features of CO_2 make it more challenging to transport in pipelines than natural gas, e.g. a greater susceptibility to long-running ductile fracture propagation [2]: a greater likelihood of lower temperatures and reduced toughness due to the Joule-Thomson cooling effect (−20 °C for line venting and down to −80 °C for leakage): and a high potential corrosion rate if an aqueous phase is present.

CO_2 has been transported and used in the food industry and for undertaking enhanced oil recovery (EOR) for decades. Large-scale transport of CO_2 is, therefore, not a new technology. More than 5000 km of dense phase CO_2 pipelines have been or are in operation worldwide, mainly constructed from carbon and low alloy steels (CLASs). The majority of the CO_2 pipelines are located in North America, where there is over 35 years' experience in carrying CO_2 from mostly natural sources to oilfields as part of EOR operations through an extensive CO_2 pipeline infrastructure [3]. Several publications deal with corrosion issues related to carbon capture and storage (CCS) [4, 5]. No serious corrosion problems have been reported in the part of the system that has been exposed to reasonably dry and pure CO_2. According to OPS statistics, there were only 12 leaks from CO_2 pipelines reported from 1986 through 2006 – none resulting in injuries to people [6].

The good experience with CO_2 transport in the USA by means of a CLAS pipeline network is often referenced to argue that CO_2 pipeline transport will not be a major challenge for CCS. The justification for this view can be questioned as CO_2 captured from fossil-fuelled power plants and other industrial sources might give dense phase

CO_2 containing impurities, that have not been transported in the past. It is also regarded as more challenging to operate a CO_2 network with many point sources and to transport CO_2 to offshore storage sites.

12.2 CO₂ Stream Composition

The flue gas from power plants and the CO_2 released from steel and cement production contain a variety of components that might partly follow the CO_2 stream through the capture and compression processes. These components, referred to as impurities in the CO_2 stream, might affect the flow properties, corrosion, and injectivity in the reservoir. If the remaining impurity concentration in the compressed CO_2 is too high, additional cleaning is required. Additional cleaning and post processing tasks add to the cost and need to be minimised/optimised.

A number of CO_2 specifications and recommendations for maximum impurity concentrations have been published, examples of which are included in Table 12.1. The CO_2 quality recommendation that has been most cited has been suggested in the DYNAMIS project [7]. Other frequently referenced CO_2 specifications have been presented by IPCC [3] and Kinder Morgan [14]. In 2012 and 2013, the National Energy Technology Laboratory (NETL) issued Quality Guidelines, giving recommendations for the impurity limits to be used for conceptual design of carbon steel pipelines [8, 9]. The recommendations were based on a review of 55 CO_2 specifications found in the literature. The most recent (2016) recommendation from 'The Carbon Net project' [10] is also included in Table 12.1.

It is apparent from Table 12.1 that there is a large variation in the reported impurity concentrations. This is considered reasonable as the impurities in the CCS or carbon capture use and storage (CCUS) streams will depend on the fuel type, the energy conversion process (post-combustion, pre-combustion, or oxyfuel) and the capture process. In addition, with new capturing technologies, new compounds (impurities) can be formed and higher concentrations of impurities can follow the CO_2 stream with an unknown effect on corrosion and cross-chemical reactions in the bulk phase.

The justification for many of the proposed recommendations can be questioned as the reported [12, 13] CO_2 compositions presently transported in pipelines do not include flue gas impurities, such as, for instance, SO_2 and NO_2; and as concluded in a recent review [5], hardly any laboratory-backed data can be found in the literature supporting the recommended CO_2 specifications.

The lack of data was recognised in the first ISO standard for CO_2 transport that was issued in 2016 [15]. In the standard it is stated that:

> Since the maximum concentration of a single impurity will depend on the concentration of the other impurities, it is not possible due to lack of data and current understanding to state a fixed maximum concentration of a single impurity when other impurities are, or may be, present.

The standard therefore recommends consulting the most up-to-date research during pipeline design.

Table 12.1 Impurity concentrations reported in existing pipelines (CO_2 specifications recommended by Dynamis [7], NETL [8, 9], the Australian carbon net project [10] and the CO_2 specification tested in the Institute for Energy (IFE) experiment [11]).

(ppmv)	Impurity levels in existing pipelines [12, 13]				Published CO_2 recommendations [7–10]				Testing [11]
	Canyon Reef Carriers	Central Basin Pipeline	Cortez Pipeline	Weyburn	DYNAMIS [7]	NETL [8, 9]	Literature Review [8, 9]	Carbon Net [10]	IFE experiment
H_2O	122	630	630	20	500	730 [8], / 500 [9]	20–650	100	122
H_2S	<260	<26	20	9000	200	100	20–13000	100	130
CO	–	–	–	1000	2000	35	10–5000	900	0
O_2	–	<14	–	<70	<40000	40000 [8], / 10 [9]	100–40000	20000	275
NOx	–	–	–	–	100 [1]	100	20–2500	250	96
SOx	–	–	–	–	100 [1]	100	10–50000	200	69

12.3 Corrosion in the Presence of Aqueous Phases

12.3.1 Pure CO_2 and Water

A number of experimental studies with dense phase CO_2 and water have shown that the corrosion rate increases with increasing temperature, that protective $FeCO_3$ corrosion product films form when the concentration of dissolved corrosion products becomes high, and that the corrosion film can fail and give high localised corrosion rates [11, 16, 17]. The observations seem very much to follow the trends seen at lower CO_2 partial pressures in oil and gas production streams. The main difference is the much higher CO_2 pressure (around the 100 bar as indicated in Figure 12.1) in the CO_2 transport lines, giving typically a one unit lower pH, a much higher solubility of corrosion products and more H^+ ions and H_2CO_3 that can corrode the steel. The result can be extreme corrosion rates exceeding $40\,mmy^{-1}$, an example of which is shown in Figure 12.2 [18] when condensed water is present in large quantities under flowing conditions. Such situations must be avoided under all circumstances. If only minor amounts of water precipitate, the water will quickly be supersaturated with dissolved corrosion products and a much lower corrosion rate can be expected. This mechanism is similar to the case of top-of-line corrosion (TLC) and the corrosion rate will be limited by the supply of new water with low concentration of dissolved corrosion products.

12.3.2 Impurities and Formation of Corrosive Phases

When impurities, such as water, SO_2, NO, NO_2, O_2, and H_2S, are present, there are a number of complex reactions that have the potential to form sulphuric/sulphurous acid, nitric acid, and elemental sulphur [11]. The conditions under which these reactions take place are poorly understood. The CO_2 composition given in the last column in Table 12.1

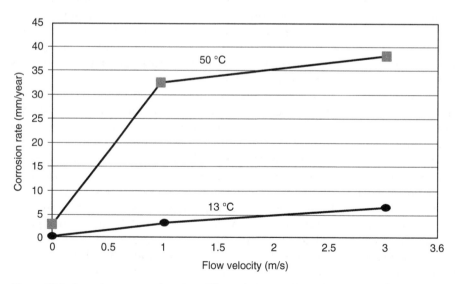

Figure 12.2 Corrosion rate as a function of flow velocity, condensed water and 100 bar CO_2. *Source:* [24].

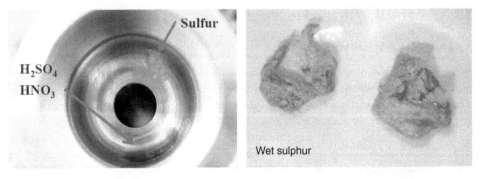

Figure 12.3 H_2SO_4, (7 M) HNO_3 (0.2 M) and elemental sulphur formed in corrosion experiment performed at 45 °C, 100 bar and CO_2 composition, as given in last column in Table 12.1, *Source:* [11]. (see colour plate section).

was tested in a rocking autoclave system at the Institute for Energy Technology (IFE) [11] and the experiment demonstrated that H_2SO_4, HNO_3, and elemental sulphur formed (see Figure 12.3) at impurity concentrations below the impurity limits given in many of the recommendations in Table 12.1.

12.4 Means of Corrosion Prediction

Avoiding the formation of corrosive phases and solids in the pipeline is essential for the safe operation of a CO_2 pipeline network. Predicting when corrosive aqueous phases form is very different from predicting the corrosion rate when such phases are present.

Water precipitates when the water solubility limit is exceeded. The solubility limit which depends on pressure, temperature, and the presence of other impurities is well known for the pure CO_2-water system, and to a certain extent for systems with small amounts of non-condensable gases, such as CH_4, O_2, N_2, and Ar. Figure 12.4 shows the water solubility in pure CO_2 at 100 bar. The shaded oblong area indicates the water concentration range specified for existing pipelines and for design of new pipeline systems. The field experience is good for the specified concentration range (20–650 ppmv), which is well below the water solubility limit. Most labs find the corrosion rates to be insignificant when the water content is below the solubility limit.

The presence of amines, methanol (MeOH), ethanol (EtOH), glycols, and other impurities giving water-soluble components like H_2SO_4 and HNO3, will facilitate the formation of an aqueous phase and reduce the concentration of water in the CO_2 at which a separate aqueous phase is formed. Experiments have shown that aqueous phases can form at water concentrations as low as 50 ppmv. If an aqueous phase forms, the corrosion rate will depend on the amount and the concentration of impurities in this phase. Conventional CO_2 corrosion prediction models outlined in Chapter 5 do not apply to CO_2 transportation under such conditions and these conventional models used by the oil and gas industry will generally over-estimate the corrosion rates

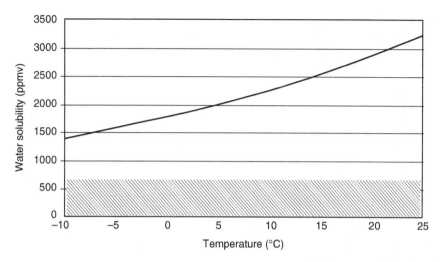

Figure 12.4 Solubility of water in 100 bar pure CO_2 as a function of temperature. The shaded area indicates the water concentration range specified for existing pipelines and for design of new pipeline systems.

12.5 Method of Corrosion Mitigation

The obvious economical choice for the transportation of CO_2 is the use of CLAS. For such a scenario, establishing safe limits of impurity levels that can be tolerated by CLAS is therefore paramount.

12.5.1 Normal Operation

The primary strategy for internal corrosion control for the effective use of CLAS is implementation of sufficient dehydration and removal of impurities in the CO_2 stream, thereby avoiding corrosive phases. The required dehydration depends on the concentration of other impurities and will be project-specific.

12.5.2 Transport of Wet CO_2

Transporting wet dense phase CO_2 is not common. The only known case is the Equinor (formerly Statoil)-operated Sleipner project where wet supercritical CO_2 is transported 12.5 km to the well head through a corrosion-resistant alloy (CRA) pipeline [19]. The pipeline is insulated in order to keep the temperature sufficiently high to prevent hydrate formation.

Transporting wet CO_2 has been considered using corrosion inhibitors. Since supercritical CO_2 is an efficient solvent for many chemicals, it has been argued that an inhibitor will partition to the CO_2 phase and become less effective. This has not been the case in laboratory studies where good inhibition has been reported [20, 21].

pH stabilisation combined with glycol has successfully been used for corrosion control in a number of gas condensate pipelines. The pH stabilisation technology has also been tested for dense phase CO_2 systems under laboratory conditions, but with limited success [20].

The use of coating for corrosion protection and flow improvement has also been considered. The main challenges are the stability against impurities in the CO$_2$ and detachment during rapid decompression giving very low temperatures (-20 to -80 °C).

12.5.3 Accidental Ingress of Water

Accidental ingress of water in a complex network of pipelines is potentially feasible. If dry CO$_2$ continues to flow after a water incident, it is assumed that the water will be dissolved quickly and not seriously threaten the integrity of the pipeline. Continuous water ingress or long-lasting shutdown after water ingress will give a quite different situation. At shutdown. it might be necessary to remove the water in the pipeline. Water removal includes depressurisation of the pipeline: experience from existing pipelines indicates that this can take weeks. The acceptable response time after water contamination will be system-specific and depend on the corrosion rate and corrosion allowance. The corrosion rate in a pipeline suffering from accidental water ingress is difficult to predict, but the worst case corrosion rate can be many mmy^{-1}.

12.5.4 Depressurisation

When dense phase CO$_2$ is depressurised and forms a two-phase gas/liquid system, impurities will divide up between the two phases and go preferentially to the phase where their solubility is highest. Experiments have shown that water, H$_2$S, and SO$_2$ accumulate, while O$_2$ is depleted in the remaining liquid CO$_2$ phase when the system is depressurised via the gas phase [16]. The experiments also showed that a separate aqueous phase could form and thus turn a non-corrosive system into a corrosive one. The corrosion rate was reasonably low (<0.1 mmy^{-1}) in pure CO$_2$ because the water phase quickly became saturated with corrosion products that reduced the corrosivity. Higher corrosion rates were encountered when other impurities were present and a higher corrosion rate persisted until the impurities were consumed. The local availability of corrosive phases, therefore, becomes important and more severe attacks can be foreseen in low spots if liquid accumulates.

More data are required in order to predict accumulation rates over a wider temperature range and for other impurities including glycol, amines, CO, and NOx.

12.5.5 Downhole Corrosion

CO$_2$ injection wells need to handle a range of CO$_2$ arrival rates within the limits of the capture plant and surface equipment. During transient (batch-wise) injection and shut-in, it is foreseen that the CO$_2$ stream and brine may occasionally mix in the bottom of the injection well and present a potential corrosion problem. Impurities present in the CO$_2$ stream will split off into the water phase at this point, water will dissolve and saturate the CO$_2$ phase, and the temperature will increase. When the tubing material is exposed to such environments, pitting corrosion and different types of cracking can become a problem.

There is not a great deal of data in the literature addressing downhole corrosion when dense phase CO$_2$ and brine are mixed [22, 23]. Key data such as partitioning of O$_2$ in water and dense phase CO$_2$ is missing. To compensate for the lack of data, a conservative approach is often applied and there is a risk of either using too expensive materials or request too strict CO_2 compositions. A very clean CO_2 was, for instance, recommended in the Peterhead project [24]. The justification for the strict CO_2 specifications has been debated and in particular the focus has been on the maximum acceptable O_2 content when 13%Cr material is used.

12.6 Summary

It is apparent that CCS is gaining significant momentum due to political and environmental aspects, as well as being a means for EOR in which transportation of CO_2 is paramount. Once captured, CO_2 needs to be transported to sites for sequestration – this is done mainly through pipelines or by tankers. Industry experience demonstrates that in the majority of conditions, CLAS pipelines can and have been used successfully and economically to transport CO_2, particularly when in 'dense phase', although this has limitations on water content and the types and amount of other contaminants. While there is no precise means of predicting corrosion in conditions containing high CO_2, further research is needed to determine such a model. Principal and key considerations in the use of a CLAS pipeline network relate to a number of elements, including transportation of wet CO_2, accidental ingress of water, occurrence of depressurisation and particular levels and types of impurity. The subject is by no means exhausted and in definite need of far more exploratory research and development.

References

1 IEA, Energy technology perspectives, scenarios and strategies to 2050, 2010.
2 A Cosham, DG Jones, K Armstrong, et al. Analysis of two dense phase carbon dioxide full-scale fracture propagation tests. 10th International Pipeline Conference, IPC2014, Calgary, Canada, http://dx.doi.org/10.1115/IPC2014-33080, 2014.
3 IPCC (2005). *IPCC Special Report on Carbon Dioxide Capture and Storage. Prepared by Working Group III of the Intergovernmental Panel on Climate Change.* Cambridge: Cambridge University Press.
4 Kermani, B. (2013). *Carbon Capture, Transportation, and Storage (CCTS): Aspects of Corrosion and Materials.* NACE International.
5 Halseid, M., Dugstad, A., and Morland, B.H. (2014). Corrosion and bulk phase reactions in CO_2 transport pipelines with impurities: review of recent published studies. *Energy Procedia* 63: 2557–2569.
6 Parfomak, P.W. and Folger, P. (2008). *Carbon Dioxide (CO_2) Pipelines for Carbon Sequestration: Emerging Policy Issues.* CRS.
7 De Visser, E., Hendriks, C., Barrio, M. et al. (2008). Dynamis CO_2 quality recommendations. *International Journal of Greenhouse Gas Control*, 2: 478–484.
8 NETL, Quality Guidelines for Energy System Studies, CO_2 impurity design parameters, DOE/NETL-341/011212, 2012.
9 NETL, Quality Guidelines for Energy System Studies, CO_2 impurity design parameters, DOE/NETL-341/011212, 2013.

10 The Carbon Net Project, Development of a CO_2 specification for a CCS hub network Project No. 2269886A-PWR-REP-001 Rev04, 2016.

11 Dugstad, A., Halseid, M., and Morland, B.H. (2014). Testing of CO_2 specifications with respect to corrosion and bulk phase reactions. *Energy Procedia* 63: 2547–2556.

12 Mohitpour, M., Seevam, P., Botros, K.K. et al. (2012). *Pipeline Transportation of Carbon Dioxide Containing Impurities*. ASME.

13 A Oosterkamp and J Ramsen, State-of-the-art overview of CO_2 pipeline transport with relevance to offshore pipelines. Open Polytec report: POL-O-2007-138-A, 2007.

14 K Havens, Kinder Morgan presentation at the Indian Center for Coal Technology Research. 2013. http://verden.abcsok.no/index.html?q=Kinder%20Morgan%20 water%20specification%20Havens-CCTR-June08%5B1%5D&cs=latin1 (accessed 7 October 2018).

15 ISO, Carbon dioxide capture, transportation, and geological storage/ ISO/TC 265, 2011.

16 A Dugstad, M Halseid, and AO Sivertsen, Dense phase CO_2 corrosion and the impact of depressurization and accumulation of impurities. NACE Annual Corrosion Conference, Paper No. 2785, 2013.

17 Dugstad, A., Morland, B.H., and Clausen, S. (2011). Corrosion of transport pipelines for CO_2 – effect of water ingress. *Energy Procedia* 4: 3063–3070.

18 Heddle, G., Herzog, H., and Klett, M. (2003). *The Economics of CO₂ Storage*. LFEE 2003–003 RP. MIT Press.

19 Eiken, O., Ringrose, P., Hermanrud, C. et al. (2013). Lessons learned from 14 years of CCS operations: Sleipner, In Salah and Snøhvit. *Energy Procedia* 4: 5541–5548.

20 S M Hesjevik, S Olsen, and M Seiersten Corrosion at high CO_2 pressure. NACE Annual Corrosion Conference, Paper No. 3345, 2003.

21 Y Zhang, K Gao, and G Schmitt, Inhibition of steel corrosion under aqueous supercritical CO_2 conditions. NACE Annual Corrosion Conference, Paper No. 11379, 2011.

22 Svenningsen, G., Morland, B.H., Dugstad, A., and Thomas, B. (2017). Stress corrosion cracking testing of 13Cr stainless steel in dense phase CO_2 with oxygen. *Energy Procedia* 114: 6778–6799.

23 B Kermani and F Dagurre, Materials optimization for CO_2 transportation in CO_2 capture and storage. NACE Annual Corrosion Conference, Paper No. 10334, 2010.

24 Shell UK Limited, Peterhead CCS Project, Well Technical Specification rev. K03, 2015.

13

Corrosion Under Insulation (CUI)

Corrosion under insulation (CUI) continues to be a major common challenge on a worldwide basis that is shared by all the refining, petrochemical, power, industrial, onshore, and offshore industries. It is not a new corrosion threat, although it can become a serious problem. CUI has been responsible for many major leaks with significant consequences in terms of health and safety incidents, discharge to the environment, lost/deferred production, and large maintenance budgets required to mitigate the problem.

CUI refers to the external corrosion of piping and vessels fabricated from carbon manganese, low alloy, and austenitic stainless steels. It occurs underneath externally clad/jacketed insulation due to the ingress of water. By its very nature, CUI tends to remain undetected and the damage does not become evident until the insulation and cladding/jacketing are removed to allow inspection or when leaks occur. Also the visual presence of heavy rust staining of the insulation can be treated as an indicator, although not necessarily indicating the actual location of corrosion. In addition, detection of the presence of free/trapped water and wet insulation – e.g. through infrared (IR) thermography or neutron backscatter – can be taken as indicative of CUI being present as a credible threat.

CUI manifests itself in many forms including general, localised, or stress corrosion cracking (SCC) depending on the materials of construction. This chapter summarises the historical background to this type of corrosion threat, outlines precautionary and design measures to minimise its occurrence and also outlines methods available for its mitigation. The chapter is not intended to be exhaustive and only outlines key parameters and avenues. It includes a section appraising a CUI prevention strategy to provide long-term and reliable prevention of CUI, moving towards an inspection-free, maintenance-free, operating mode that can significantly reduce the need for, and cost of, piping maintenance.

13.1 Historical Context

The battle to combat CUI had been fought for many years in the petrochemical industry, but it was perhaps the publication of ASTM STP 880 [1] that marked the modern CUI battleground. This ASTM publication reviewed the causes and factors affecting the

Corrosion and Materials in Hydrocarbon Production: A Compendium of Operational and Engineering Aspects, First Edition. Bijan Kermani and Don Harrop.
© 2019 John Wiley & Sons Ltd. This Work is a co-publication between John Wiley & Sons Ltd and ASME Press.

(a) (b)

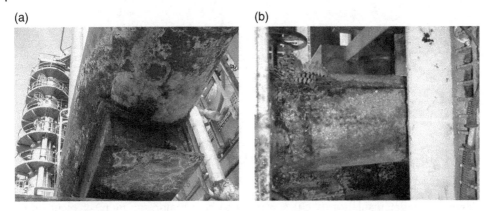

Figure 13.1 (a) and (b) typical examples of CUI showing devastating consequence of the corrosion threat. (see colour plate section).

occurrence and rate of CUI, the field experience with insulation types and control measures, including the use of coatings, specifications, system design, and inspection.

Since then, a number of events, conferences, and international forums have been held to discuss the cause of CUI, its consequences, and mitigating methods which have led to the compilation of pertinent mechanistic and practical documents [2–5].

13.1.1 Key Features

As mentioned, CUI can take many forms, depending on the metallic material deployed. In carbon manganese and low alloy steels, significant general or localised corrosion is the most common mode of failure.

CUI also occurs in austenitic stainless steels in which the type of damage is pitting corrosion, although the most common problem is chloride-related SCC without any significant loss of metal. Typical examples of the threat presented by CUI are shown in Figures 13.1a and 13.1b, some with devastating consequences.

13.2 Key Parameters Affecting CUI

CUI usually occurs when both water or moisture and oxygen are present and are in contact with steel substrate, allowing oxygen corrosion to occur. Water ingress can be due to a number of reasons, including breaks in the insulation cladding/jacketing material which may have resulted because of poor installation, damage during service or simply because of deterioration over time. This section outlines the key parameters that influence the occurrence of CUI, as summarised in Figure 13.2.

13.2.1 Water

The principal sources of water are external in origin and include rainwater, deluge systems, cooling tower drift, process and cooling water leaks, and condensation. This water may be retained, depending on the absorption properties of the

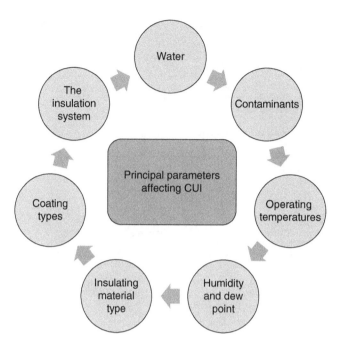

Figure 13.2 Principal parameters influencing the occurrence of CUI.

insulation material and the operating temperature. Depending upon the process conditions, saturated insulation may never dry out completely. Wet/dry cycles can occur at higher operating temperatures which can lead to exacerbation in concentration of any contaminants that may be present in the water. These elements are schematically portrayed in Figure 13.3.

13.2.2 Contaminants

Contaminants are essential components causing CUI on both carbon manganese and carbon and low alloy steels (CLASs) as well as austenitic stainless steels.

The presence of oxygen provides a ready cathodic reactant for the resultant corrosion cell established beneath insulation. The cathodic kinetics may be further exacerbated by the presence of pollutant acid gases, such as CO_2 and NO_x, depressing the water's pH (more acid) as it will have poor pH-buffering capacity. However, chlorides and sulphides make up the bulk of the contamination that generally increases significantly the corrosivity of the water and affects the resulting form of attack. The source of the contaminants can be external, such as environmental-borne chloride sources from marine environments (e.g. offshore), or windborne salts from cooling tower drift, or from periodic testing of firewater deluge systems. Contaminates can also be produced by leaching from the insulation material itself. The presence of an applied or residual stress and temperatures exceeding 60 °C (140 °F) and high chloride contents of water can contribute to external chloride SCC (E-Cl⁻SCC).

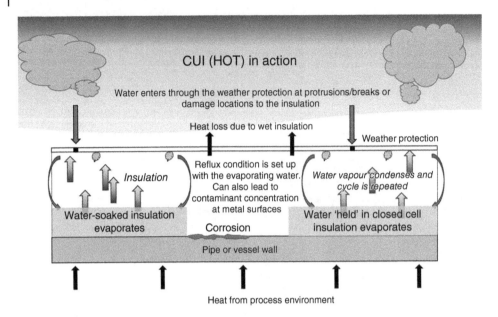

Figure 13.3 Sources of water in causing CUI.

13.2.3 Primary CUI Temperature Ranges

The accepted operating temperature range that usually causes susceptibility to CUI is within −4°C to 175 °C. This is the actual metal temperature which may not reflect the perceived operating or design temperature. This temperature range reflects the experience from the chemical process industries (CPI), oil and gas and related industries.

However, CUI has been reported at temperatures as low as −40 °C (and below) and above 500 °C due to the presence of hot and cold 'fingers' at locations where the insulation has broken down or has deteriorated during service. Equipment that has 'dead legs' (hot or cold with no process flow) or equipment subject to temperature extremes or cycling during normal operation or start-up/shut-downs is also susceptible. For example, CLAS cryogenic equipment cycling from 'cold' to ambient temperatures is subject to CUI at breaks in the insulation; or stainless steel equipment cycling from high temperatures through a temperature range between 60 and 175 °C can be susceptible to E-Cl⁻SCC and equipment in sweating environments (operating below the dew point).

13.2.4 The Effect of Temperature on CUI

The service operating temperature is an important key parameter affecting CUI. Figure 13.4 [6] shows the effect of increasing temperature on corrosion of CLASs and introduces the concept of a closed system in which oxygenated water evaporation is limited, hence resulting in reducing the corrosion rates. Two factors are involved in such a system:

- a higher temperature reduces the time that the metal surface is wet;
- a higher temperature tends to increase the corrosion rate and reduce the life of protective coatings, sealants, etc.

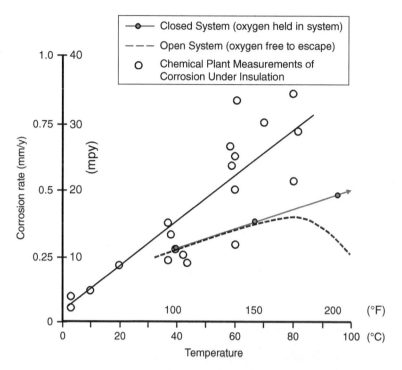

Figure 13.4 The effect of operating temperatures on corrosion rate of carbon and low alloy steels in oxygenated water systems. *Source:* [6].

Overall, increasing service temperature results in accelerated corrosion at rates higher than anticipated and explains the reason behind the significance of CUI becoming a serious problem.

13.2.5 The Effect of Humidity and the Dew Point: Sweating Corrosion

It is widely known that the corrosion rate of CUI is normally higher at coastal locations or in 'hotter' countries (e.g. the Gulf Coast, Far East Asia) than in the Northern Europe or land-locked locations. This is especially pertinent for equipment operating close to freezing temperatures (below the ambient temperatures) or in sweating service operating below the dew point (DP). The primary difference between the locations is the average temperature and the relative humidity levels throughout the year. Figure 13.5 shows DP in terms of location and month indicating that when temperature variation is large, CUI can become more likely. The humidity and operating temperature control the DP and the DP controls the degree of wetness; and when the temperature is high and does not vary, the potential for CUI is more likely. A typical example of sweating CUI damage accelerated by galvanic corrosion is shown in Figure 13.6.

13.2.6 The Effect of Insulation Type on CUI

Both closed and open cell insulation materials have been associated with CUI. No insulation system is immune from CUI, contrary to claims made by insulation suppliers. An example of CUI of piping insulated with expanded pearlite type insulation is shown in Figure 13.7.

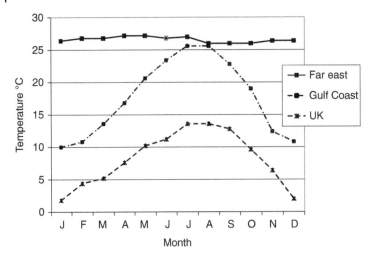

Figure 13.5 Dew point (DP) temperature in terms of location and month.

Figure 13.6 A typical example of carbon steel sweating CUI damage accelerated by galvanic corrosion at a carbon steel/stainless steel flange material specification change. (see colour plate section).

Figure 13.7 Isolated CUI of piping with expanded pearlite insulation. (see colour plate section).

Open cell insulation materials, such as mineral wool, foamed in place polyurethane (FIPP) or calcium silicate can absorb water/moisture, which has the effect of prolonging the time of wetting in a system that is intermittently wet. Closed cell insulation materials, such as cellular glass, aerogels, and expanded pearlite or expanded polyolefin do not absorb water but do trap water.

The installation procedures for all types of insulation are key to determining the effectiveness as an insulating material and resistance to CUI.

13.2.7 The Insulation System

Insulation is normally considered as a system comprised of three separate components:

- a protective coating;
- insulation material;
- external cladding.

CUI can occur due to failure of one or all components of the insulation system as follows:

- external cladding failure, allowing water to penetrate the insulation;
- accumulation of water and contaminants within the insulation;
- failure of the protective coating, allowing contact of moisture/water with the substrate, hence initiating corrosion.

13.3 CUI Prevention Methods

Current hydrocarbon industry CUI management plans include all or most of the following elements:

- Correct insulation system design and installation to exclude/prevent water penetrating the insulation system.
- Correct design of insulation sheathing so that water runs off the insulation.

- Improving insulation system designs, especially on horizontal sections or at changes in the direction where openings in the insulation often occur.
- The use of insulation that does not retain moisture, examples include cellular glass or pearlite in place of commonly used fibreglass.
- The use of insulation that is free of potentially harmful substances, such as exclusion of chlorides in fibrous insulation.
- Allowing the equipment to run without insulation in which case there is a need to consider heat losses and personnel protection before adopting this method.

A number of these elements are described briefly below.

13.3.1 Protective Coatings

Use of a conventional paint or organic coatings on steel is a matter of choice and dependent on the likelihood of CUI. Typically epoxies or epoxy phenolic coatings are used. The upper temperature limit of the coating should be appropriate for the operating service. Both thermally sprayed aluminium (TSA) and immersion-grade coatings are becoming widely used to coat and protect equipment operating in the CUI temperature range. However, in common with other coatings, TSA does have a finite life which is dependent on the initial application QA/QC, insulation system maintenance, and the operating conditions.

When the organic coating's protective life is reached, the 'out-of-sight' nature of CUI makes it difficult and expensive to detect. In such cases field re-painting is necessary to maintain a low risk of leaks. Alternatively, on-going, periodic use of a non-destructive evaluation (NDE) method with a high confidence level of detecting CUI is required to monitor the rate of CUI and quantify the piping system's remaining life. Under these conditions, the inspection costs can equal or exceed the cost of field (maintenance) painting.

13.3.2 Organic Coatings for Carbon Steel Components

Organic coatings are the primary corrosion control for CUI today, and will remain important in the future, especially for maintenance of existing piping systems or in hot work restricted areas. However, the weak points of thin-film organic coatings are their brittle nature – which leads to nicks and scratches during pipe handling and installation – and their permeability. These weaknesses are especially problematic in CUI services.

Product formulations with improved permeation resistance to increase the service life of pipeline coatings will keep the economics of organic coatings attractive. Many coating manufacturers now have new formulations specifically intended for CUI protection, with claimed upper temperature and wet heat resistance at least 100 °C (180 °F) higher than that found in coatings of a few years ago. Continued development and evaluation of organic coatings remain an important contribution to CUI prevention technology.

Table 13.1 summarises the key attributes and shortfalls of each coating system in terms of mitigating CUI.

Table 13.1 Comparison of thermally sprayed aluminium (TSA) and conventional paint in terms of CUI capabilities (online field application comparison).

Features	TSA	Conventional paint
CUI protection	25–30 years, maintenance-free and inspection-free	5–13 years, tends to low side for in-line application
Protection in cyclic service	Yes	No effective paint system
Upper continuous operating temperature	480 °C (if a seal coat is not applied)	Typically 175 °C (up to 540 °C with specialist paint systems)
Schedule impact	None – one-coat application (if a seal coat is applied, then same cure required as for the paint)	24 hours typically, multiple coats required
Environmental Impact	None (for seal coat, same as paint)	Must meet VOC and disposal regulations
In-place cost ratio	1.05–1.20	1.0
Durability	Very resistant to mechanical abuse. Minor damage does not result in CUI.	Very susceptible to mechanical abuse. Any damage is likely to result in CUI
Required surface preparation	White/near white (SA $2^1/_2$)	White/near white (SA $2^1/_2$)
Application method (s)	Twin arc spray or flame spray (for seal coat – as for paint)	Spray, brush or roller
Application accessibility	Arc/spray head to within an angle of 300 normal to surface	Brush/roller for restricted access but life decreases
Application temperature limit	None but service must be dry unless applying seal coat	Ambient to about 60 °C
Work permit required	Hot work	Cold work, but can restrict hot work in the area where painting is taking place

13.3.3 Thermally Spray Aluminium (TSA)

TSA application by electric arc or flame spray has been described extensively in the available literature. TSA has provided atmospheric corrosion protection for over 40 years on structures such as bridges, locks and penstocks. These experiences have been standardised in DOD-STD 2138 [7]. This development effort established that TSA can provide long-term protection in severe CUI environments with significant life-cycle cost savings. Initial costs, however, have been higher than organic coatings, and this has slowed the spread of TSA to other industries. More recently, the development of equipment with higher deposition efficiency and greater mobility has helped reduce the initial costs and increase market penetration, especially in the petrochemical industry.

The main advantages of TSA coatings over conventional organic coatings include:

- Longer life expectancy with minimal requirements for maintenance and inspection. Resistance to mechanical damage.
- No drying/curing time required after application – can be used immediately.

- Greater range of temperature resistance than organic coatings (−100 to 500 °C).
- Provides sacrificial protection to steels in aqueous environments.

The main disadvantages of TSA coatings over conventional organic coating include:

- Possibly higher cost of application.
- Increased difficulty of field application.
- Resistance to change by operations and maintenance organisations.

13.4 CUI Mitigation Strategy

There is a fundamental realisation that the maintenance portion of the systems-specific approach needs to be optimised by concentrating on more fundamental prevention methods, as opposed to mitigation and periodic renewal.

Maintenance procedures for insulation systems that do not use TSA are expensive and require periodic stripping, abrasive blasting, recoating, and re-insulating; and/or on-going periodic NDE/inspection activities. Insulation systems that do use TSA are characterised by their ability to provide longer-term CUI prevention and have a lower failure potential over their longer life-cycle and are, therefore, not as dependent on maintenance and inspection activity to manage CUI.

Once the protective life of the organic coating is reached, the CUI prevention measures discussed below are based on practices common in the petrochemical as well as other industries. The following CUI prevention or mitigation strategies have been developed to prolong the life of all insulated equipment. They maintain a lower failure potential over a longer life-cycle and are, therefore, not as dependent on the effective but expensive maintenance and inspection activities that are required to manage CUI:

- Upgrading to stainless steel metallurgy when economically justified.
- Removing unnecessary insulation.
- Replacing insulation at pipe support vents, etc. with a seal either side to remove the water ingress point.
- Use of aluminium foil to prevent external chloride SCC (E-Cl⁻SCC).
- Use of true waterproof/impervious non-metallic weather protection barriers.

13.4.1 Stainless Steel for Small Diameter Piping

Small diameter piping – 3″ nominal pipe size (NPS) or less – appears to be prone to CUI leaks because of its low wall thickness, the increased number of field welds, the coating's inefficiency, and the human tendency to pay less attention during handling, maintenance, and inspection. Stainless steel piping would solve the CUI concerns in many services but the initial cost and possibility of external stress corrosion cracking or pitting have been impediments to wider use. However, even at today's prices and at equal schedules, the life-cycle cost savings for stainless steel piping versus painted carbon steel can be significant. Small diameter stainless steel piping has a role to play in selected applications to prevent CUI.

13.4.2 Aluminium Foil Wrapping

Aluminium (Al) foil wrapping now presents a low-cost, proven option for preventing external stress corrosion cracking and pitting, leaving the initial cost issue as the outstanding item. The Al foil provides cathodic protection (CP) by acting as a sacrificial anode in the same manner as TSA. The potential required to mitigate the Cl⁻SCC of austenitic stainless steel using Al foil is much lower than that required to mitigate CUI on CLAS using TSA, resulting in long-lasting protection in more instances.

The use of Al foil has a proven track record and is recommended for all austenitic stainless steel equipment and piping which may be prone to CUI. Aluminium foil can also be used on austenitic stainless steel vessels in lieu of conventional coating for E-Cl⁻SCC cracking protection – costs should be economically based.

Al foil should not be used in sweating service because the service life of Al foil will be shortened. TSA should be considered favourably in sweating service. Concern over possible liquid metal cracking of austenitic stainless steels has been expressed; however, operating experience to-date and literature searches have not highlighted any instances of Al initiating liquid metal cracking.

Cost comparisons in both North America and Europe showed that the cost of Al-foil wrapping was 60–80% of the conventional coating cost. A comparison between Al foil and conventional paints in terms of CUI capabilities is given in Table 13.2.

13.4.3 Remove Unnecessary Insulation: Personnel Protection Cages

Thermal insulation is used to protect workers from hot surfaces and conserve energy. In services where the thermal insulation is applied only for personnel protection, wire 'stand-off' cages can replace the insulation. These cages are simple in design, low in cost, and free of concerns with CUI. Typical examples of cages are shown in Figure 13.8. The initial cost of personnel protection cages is 5–15% less than the installed cost of thermal insulation, and, again, life-cycle costs are a bit lower than those for organic coatings. Personnel protection cages have been used in the petrochemical industry for over 30 years. Their use now appears to be growing as companies continue to explore ways of reducing initial and on-going maintenance costs.

13.5 CUI Inspection

Risk-based inspection (RBI) is widely accepted and established in the refining, petrochemical and offshore industries. CUI inspection should follow a similar approach. The RBI assessment makes use of actual operational and structural conditions of insulated systems, not the design conditions. In order to obtain valid information from the RBI analysis, it is important to be sure that all the input data are correct.

When introducing the RBI approach, insulated systems must be assessed to determine the appropriate risk levels to determine inspection plans. On large refining and petrochemical sites, it will be impossible to conduct such an effort on all insulated systems at once, because of limited resources and budget. For that reason, a high level prioritisation step process (Figure 13.9) on a unit-by-unit basis where insulation is

Table 13.2 Comparison of Al foil and conventional paint in terms of CUI attributes

Feature	Aluminium foil	Conventional paint
Corrosion protection for equipment operating in ambient to moderate temperature range	25–30 years based on ICI reported experience	9–13 years maximum depending on environment (dry)
High temperature > 190 °C (350 °F) – cyclic service corrosion protection	No decrease in service life	No effective paint system exists
Upper temperature limit	540 °C (1000 °F) dry	150–230 °C (300–450 °F)
Chemical resistance	Resistant to all solvents, but narrower pH resistance range (not resistant to strong acids or bases)	Wide pH resistance range, but not resistant to solvents
Cure time between coatings	None	Approximately 24 hours between coatings
Environmental impact	None	Must meet VOC and disposal regulations
Application cost for piping (painted carbon manganese steel equal to 1)	2 Nominal Pipe Size (NPS) = 1.26 4 NPS = 1.54 8 NPS = 2.69	2 NPS = 2.07 4 NPS = 2.76 8 NPS = 4.79
Durability.	Excellent. Minor damage will not result in corrosion.	Very susceptible to mechanical abuse. Any damage to coating will result in corrosion.
Required surface preparation	None	SA 2½ or better.
Application method(s)	Overlapping wrap of Al-foil	Spray, brush, and roller
Application accessibility	Same as for insulation	Able to apply to surfaces with restricted access using brushes and rollers
Work permit required	Cold work	Cold work, but it can restrict hot work in the area where painting is taking place

 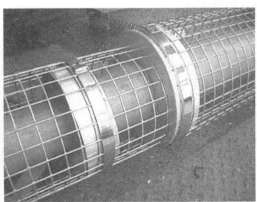

Figure 13.8 Typical examples of cages as a replacement for insulating materials.

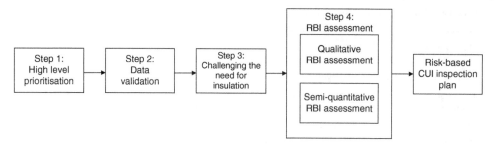

Figure 13.9 Typical steps required to develop a risk-based CUI inspection plan.

present is introduced that may help to prioritise the RBI efforts. Using this approach, one will generally be able to initially direct RBI efforts to those insulated systems that feature the highest risks for operations.

Once the process units have been prioritised with respect to risk of CUI failure, it is recommended to carefully challenge the need for insulation. It is clear that the best way to eliminate CUI is to eliminate insulation. Figure 13.9 shows the steps required to develop a risk-based CUI inspection plan. In addition, factors that need to be considered when addressing the probability of failure (POF) are outlined in Table 13.3.

The steps that are necessary for a risk-based screening exercise include:

1) Obtain commitment from management.
2) Assemble a multi-disciplinary RBI team.
3) Identify the POF and consequences of failure (COF) criteria.
4) Identify the areas of applicability.
5) Gather all applicable data and information (process operating, equipment history, specifications, etc.).
6) Determine the timeframe for the assessment.
7) Define the failure criteria.
8) Develop a credible failure scenario.
9) Determine the unmitigated risks.
10) Conduct the risk assessment, i.e. POF versus COF.
11) Screen equipment with acceptable risks.
12) Develop mitigation plan for equipment with unacceptable risks.
13) Document the RBI assessment and obtain management approval.

13.6 NDE/NDT Techniques to Detect CUI

NDE/NDT techniques employed to detect and evaluate CUI can be conveniently placed in one of two categories: screening or direct assessment techniques. These are briefly described in Table 13.4. Each individual technique has both advantages and disadvantages, but no single technique can be relied on to provide full confidence. The most effective inspection technique is to completely remove the insulation and carry out a full external visual inspection. This is probably the most expensive option. Most CUI Inspection plans involve a combination of RBI, screening, and direct assessment techniques together with an element of CUI mitigation strategy.

Table 13.3 Factors to consider when determining CUI probability of failure (POF)

Data required	Explanation
Coating information	TSA/organic coating/no coating, age of coating
External environment	Environmental conditions (climatic), cooling tower drift, sweating and dripping water, deluge, steam trace leaks
Operating temperature	Establish if metal temperature is in the CUI range −4–149 °C (25–300 °F), and whether in the range 60–100 °C (140–212 °F)
Metallurgy	Carbon and low alloy steels, stainless steels or other alloys
Insulation practices	Good/poor insulation quality control and practices (ensuring sealed cladding); coating application quality
Operating history	Constant, cyclic; temperature history
Inspection history	External visual inspection, visual inspection data after removing cladding, NDE data, repair history, time since last inspection
Equipment type	Piing or vessel; small bore piping; thin wall piping

Table 13.4 Typical NDE/NDT techniques for detecting CUI.

Screening	Direct assessment
External visual (without insulation removal)	External/visual (with insulation removal)
Flash radiography	Ultrasonic thickness (requires insulation removal or inspection ports)
Profile radiography	Guided wave ultrasonic inspection
Neutron backscatter	Digital radiography
Thermography (infra-red)	Real time radiography
Pulsed eddy current	Digital real-time radiography

13.7 Summary

CUI is a serious integrity threat that affects all the major process, chemical, oil and gas and associated industries that require widespread insulation of plant and pipework. CUI has been found beneath all types of insulation systems, coatings, and external claddings. Equipment design and the use of newer insulation materials, coatings, and cladding can help to mitigate CUI, but CUI can only be mitigated by combining the newer insulations systems with an effective Inspection and Maintenance strategy.

It is important to stress that:

- All types of insulation are prone to CUI, i.e. a change in insulation material will NOT mitigate CUI.
- All claddings fail to prevent water ingress, it is possible to minimise water ingress but impossible to mitigate it completely.

- All coating systems have a finite life. The life expectancy is dependent on the coating type, operating temperature, and installation practice. TSA is not immune to failure but affords greater protection due to the CP it provides.
- Insulation and maintenance are time-consuming and expensive.

References

1 Pollock, W.I. and Barnhart, J.M. (eds.) (1985). *Corrosion of Metals Under Thermal Insulation*. ASTM STP 880.
2 Pollock, W.I. and Steely, C.N. (eds.) (1990). *Corrosion Under Wet Thermal Insulation*. NACE International.
3 NACE (1989). *A State-of-the-Art-Report on Protective Coatings for Carbon Steel and Stainless Steel Surfaces Under Thermal Insulation and Cementitious Fireproofing*. NACE Publication No. 6H189. NACE International.
4 NACE (1998). *The Control of Corrosion Under Thermal Insulation and Fireproofing Materials: A Systems Approach*. RP 0198-98. NACE International.
5 Winnik, S. (ed.) (2008). *Corrosion Under Insulation (CUI) Guidelines*. Publication No. 55,. European Federation of Corrosion.
6 Speller, F.N. (1935). *Corrosion: Causes and Prevention*, 2e. New York: McGraw-Hill Book Co. See p. 153 and Fig. 25.
7 Department of Defense, Metal Sprayed Coating Systems for Corrosion Protection Aboard Naval Ships, DOD-STD 2138, 1981.

14

Metallic Materials Optimisation Routes

An overview of materials selection routes for different production scenarios is given in this chapter acting as a guide in a holistic strategy for materials optimisation. This provides a preferred route to the choice of appropriate materials for a particular application. A basic knowledge of materials and corrosion is nevertheless highly advisable so that a fitforservice solution is achieved. It is important to note that while carbon and low alloy steels (CLASs) are chosen primarily according to their general and localised metal loss corrosion resistance, with adequate resistance to sulphide stress cracking (SSC), corrosion resistant alloys (CRAs) are normally selected mainly based on their resistance to environmental cracking (EC). These latter threats include SSC and chloride stress corrosion cracking (Cl⁻SCC) or a combination thereof as affected by the operating temperatures and conditions. The exception for CRAs is under extreme conditions – typically a combination of high temperature, low pH, high CO_2, and H_2S – where general corrosion may also or exclusively have to be considered in the overall selection strategy. In this chapter, an integrated strategy to materials optimisation is illustrated to enable safe and trouble-free operations while maintaining economy.

14.1 Background

There is a growing desire to have a corrosion design strategy for production facilities able to handle and transport wet hydrocarbons. Such an approach can be used in the technical/commercial assessment of new field development and in prospect evaluation and to assess the risk of handling sour fluids by facilities not normally designed for sour service. Materials optimisation, therefore, together with effective whole life corrosion management remains the key operational challenge to successful hydrocarbon production, economy, and safety. In this context, selection and optimisation of appropriate materials, which can tolerate given production scenarios, are the underpinning steps.

The wide-ranging environmental conditions prevailing in oil and gas production facilities necessitate appropriate and cost-effective materials choice and corrosion control measures. The implementation of these measures is becoming increasingly important as the impact of corrosion threats on safety, economy, and the environment also takes on a challenging role. Furthermore, production conditions are tending to

Corrosion and Materials in Hydrocarbon Production: A Compendium of Operational and Engineering Aspects, First Edition. Bijan Kermani and Don Harrop.
© 2019 John Wiley & Sons Ltd. This Work is a co-publication between John Wiley & Sons Ltd and ASME Press.

become more corrosive, hence requiring a more stringent corrosion management strategy. The use of CLASs backed by a correct corrosion control system is considered most favourable for commercial reasons. The corrosion threats presented by H_2S, CO_2, and O_2 have major mechanistic and industrial implications that require a thorough understanding, as discussed in separate chapters.

This chapter briefly describes an approach to optimise materials for both production and injection systems focusing on environmental and operational parameters, bearing in mind whole life costing. Past successes in effective use of CLASs are included, highlighting key enabling criteria, allowing extended use of these alloys. A more detailed methodology describing the subject is given elsewhere [1–4] particularly in ISO 21457 [1]. First, a brief overview of the elements of production facilities is given with a view to outlining the materials selection route for each application.

14.2   Production Facilities

The primary focus of materials optimisation is normally on large capital expenditure-intensive (CAPEX) components, including well tubing/casing, flowlines, pipelines, and trunk lines. Any economic saving in such large items can have a significant impact on the commercial viability of a project. Other smaller items may constitute a lower degree of importance in terms of materials type and a conservative choice may normally be warranted as its overall economic impact may not be excessive.

14.2.1   Drilling Components

Drilling operations are accomplished by the use of drill string. Briefly, a drill string comprises a drill pipe, a heavyweight drill pipe, drill collars, and a drill bit. Each pipe is a hollow, thick-walled seamless piping, fitted with threaded ends called tool joints, that transmits drilling fluid and torque through the wellbore to the drill bit on a drilling rig.

Drill string components are primarily high strength low alloy steels, although for some specific applications, and in extended reach operations, aluminium drill pipe may be used. Corrosion threats are primarily due to drilling fluid used for well control and lubricity. Drilling fluid (mud) may contain oxygen which under prevailing operating conditions may render it corrosive. Corrosion control is, therefore, mainly implemented through the control of drilling fluid pH and, in some instances, oxygen removal.

14.2.2   Wells/Subsurface Components

The selection of materials for well completion applications includes well casing, tubing, equipment/accessories, the wellhead and Christmas tree. The choice is governed by two overriding scenarios – this is subject to direct contact with the production stream containing water and includes: (i) fluid flow wetted parts; and (ii) fluid flow non-wetted parts of the well completion. While metallurgical choice for the latter components is not affected by produced/reservoir fluid conditions and CLAS can be utilised for these parts, metallurgical solutions for the former are governed by produced fluid and the operating conditions. Therefore, the choice of materials for fluid flow wetted parts is typically governed by the need for resistance to both metal loss corrosion and to aspects of cracking/environmental cracking (EC). EC is important even at low levels of H_2S under prevailing in-situ pH conditions due to the high pressure and/or high concentration of chloride and the need for long-term reliability to avoid potential safety risks and unnecessary workover costs.

Use of CLASs together with the application of downhole inhibition for low to moderate corrosive conditions has proved to be somewhat successful in certain conditions, particularly for onshore wells. This is subject to the prevailing operating conditions and in particular acid gas contents, as outlined in Chapter 5. However, such systems can often prove impractical or too costly (e.g. deepwater subsea developments), so that they are not always the best approach. For highly corrosive conditions, CRAs remain the most effective and economic option.

14.2.3   Manifolds

A production manifold is a large subsea/onshore structure made up of various valves and pipework, designed to commingle and direct produced fluids from multiple wellheads into one or more flowlines. Manifolds are usually mounted on a template and often have a protective structure covering them. Due to the complex and critical nature of such structures, they are more often than not manufactured from CLASs with CRA cladding on the fluid flow wetted parts. They are difficult to inspect and, therefore, should be designed for the design life of the field/reservoir. A brief summary of potential options and materials of construction for manifolds is shown in Table 14.1.

14.2.4   Flowlines and Unprocessed Fluids Pipelines

The choice of materials and corrosion mitigation methods for in-field flowlines and unprocessed fluid pipelines is considered by two scenarios, based on perceived system corrosivity and respective potential risks:

i)   *Highly corrosive or high-risk applications:* For highly corrosive or high-risk conditions, CRAs or internally clad CRAs often remain the most cost-effective and

Table 14.1 Potential CRA options for manifolds.

Alloy	Advantages	Disadvantages
13%Cr SS	• Lowest material cost • High strength	• Very low H_2S tolerance • Availability of fittings • Potential hydrogen embrittlement from external CP
22%Cr duplex SS	• Good availability • High strength	• Low H_2S tolerance • Potential hydrogen embrittlement from external CP
25%Cr super-duplex SS	• Corrosion resistance • Very high strength	• Low H_2S tolerance • Complex heat treatment • Potential hydrogen embrittlement from external CP
CS internally clad (625, 825)	• Greatest corrosion / H_2S resistance • No H_2 embrittlement	• Fabrication complexity • Availability of fittings

reliable option. In such conditions, the associated consequence of failure can be high and, therefore, the use of corrosion inhibition with CLASs can often either be impractical, costly or pose too high a risk. A brief summary of potential options and materials of construction for pipelines is shown in Table 14.2.

ii) *Low to moderate corrosiveness or low-risk applications:* In low to moderate corrosive and low-risk conditions, CLAS with corrosion inhibition or pH stabilisation is an effective option. Corrosion inhibition is normally implemented by continuous injection. The corrosion inhibitor must be selected appropriately in accordance with field conditions and operating parameters, as discussed in Chapter 8. In such situations, on-line corrosion monitoring and periodic mechanical and intelligence pigging may be required to ensure effective inhibition and, as important, inhibitor replenishments particularly where deposits may drop or accumulate within the flowlines.

Non-metallic liners using high or medium density polyethylene (M or HDPE) and special grades of Nylon (Rilsan) have been used successfully to reduce failures and inhibition cost, although the limits of applicability and operational deployment and long-term durability for such liners need to be taken into account.

Table 14.2 Subsea CRA options for pipelines

Alloy	Advantages	Disadvantages
13%Cr SS	<ul><li>Lowest material cost</li><li>High strength</li><li>Less reduction in strength with increasing temperature</li></ul>	<ul><li>Very low H_2S tolerance</li><li>Availability of fittings</li><li>Potential hydrogen embrittlement from external CP</li><li>Welding/PWHT[a]</li><li>Sea water ingress corrosion</li></ul>
22%Cr duplex SS	<ul><li>Good availability</li><li>High yield strength</li><li>Corrosion resistance</li></ul>	<ul><li>Low H_2S tolerance</li><li>Potential hydrogen embrittlement from external CP</li><li>Reduction of yield with increasing temperature</li><li>Sea water ingress corrosion</li></ul>
25%Cr super-duplex SS	<ul><li>Corrosion resistance</li><li>Sea water tolerant</li><li>Very high strength</li></ul>	<ul><li>Low H_2S tolerance</li><li>Complex heat treatment</li><li>Potential hydrogen embrittlement from external CP</li></ul>
CLAS internally lined/clad (316L, 825, 625)	<ul><li>Optimised corrosion/H_2S resistance</li><li>No H_2 embrittlement</li><li>Can use high strength CS line pipe grade</li></ul>	<ul><li>Fabrication complexity</li><li>Limited availability</li><li>Lined pipe cannot be reeled</li><li>Sea water ingress corrosion of low PREN CRAs</li></ul>

[a] Potential delay in installation when post weld heat treatment (PWHT) is required, causing increased cost.

14.2.5 Flexible Pipes

The topic of flexible pipes is dealt with in Chapter 15 and no further description is necessary here.

14.2.6 Process/Surface Facilities

In general, process facilities will require the same materials selection and corrosion mitigation strategy as the in-field flowlines. In addition, vessels may require internal corrosion barriers (briefly discussed in Chapter 9) to prevent under-deposit corrosion and/or corrosion inhibition to mitigate the potential corrosion of CLAS components. Selection of organic coatings versus CRA cladding will depend on the corrosivity of the processed fluids, the presence of solids or when there is potential impact of H_2S presence on degradation of organic coatings due to blistering, etc. It should be said that in process facilities, deployment of a corrosion inhibitor is more simply achieved subject to the complexity of the facilities, such as piping geometry, bends, local flow conditions, dead legs, and others.

Heat exchangers, particularly gas coolers, will require CRA material as a minimum for the heat exchanger tubes and heads or shell, depending on the design of the unit and system corrosivity.

Process piping can be either CRAs or CLASs with corrosion inhibition (depending upon the piping configuration, quantities, system corrosivity, etc.), except for high temperature areas (>100 °C) where CRAs are more suitable. Non-metallic materials, such as HDPE, FRP and lined piping, are practical alternatives for produced water handling and are becoming more widely used.

14.2.7 Gas Treating Plants

CLASs with continuous corrosion inhibition are normally acceptable options for gas treating plants. The exceptions are the high corrosive sections of the plants, such as amine towers, glycol reboilers, and gas coolers, where cladding or solid CRAs may be required.

14.2.8 Export Pipelines and Trunklines

Export pipelines and trunk lines are CAPEX-intensive and, therefore, subject to system corrosivity, normally require the use of inhibited CLASs to mitigate any corrosion that may occur.

14.2.9 Seals and Elastomers

Elastomers and non-metallic materials are briefly dealt with in Chapter 15. It should be noted that H_2S, even at low concentrations, can cause severe degradation of elastomers and increases the risk of elastomer embrittlement and explosive decompression. Selection of seals and elastomers for H_2S/CO_2 environments must follow the same selection guidelines for pure H_2S environments.

14.3 The Operating Regimes

The choice of material is governed by the nature of its application and generally falls into two broad categories of production and injection. Accordingly, selection can be made on the basis of appropriateness, corrosivity evaluation, and economy. The process by which material optimisation is achieved is governed by due consideration of several parameters, as summarised in Figure 14.1 and described in Section 14.10. Typical material systems used for these applications are summarised in Figure 14.2.

14.3.1 Production

Selection of metallic materials for production duties in which three phase oil, gas, and water may be present can be made, using a number of options referred to in Figure 14.2. Subject to system corrosivity, CLASs have offered satisfactory performance, albeit a growing number of CRAs or non-metallic options are being used.

14.3.2 Injection

As described in Chapter 7, water and/or gas injection is implemented to maintain/increase the reservoir pressure or discard the produced water or gas to remove the

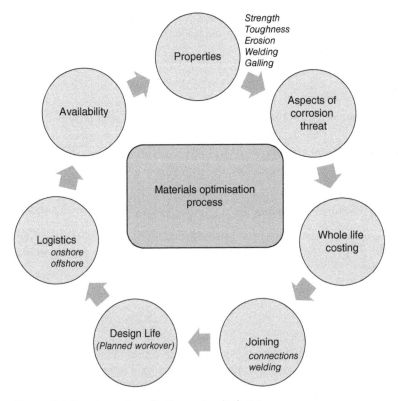

Figure 14.1 Key parameters affecting materials choice.

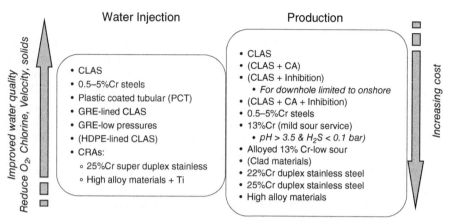

(Those in brackets are generally applicable to flowlines and pipelines)

Figure 14.2 Principal choice of alloys used in CAPEX intensive production and injection systems. Manifolds (and possibly wellheads) and risers may be manufactured from CRA internally clad. CA = corrosion allowance; GRE = glass reinforced epoxy; HDPE = high density polyethylene.

environmental impact. Over the years, subject to maintaining water quality or fully dried gas, CLASs have provided satisfactory performance. However, the use of CRAs may prove beneficial in situations where water quality cannot be maintained, e.g. where satisfactory oxygen removal may not be feasible, oxygen excursions may be foreseen or for untreated water injection. In these situations, only fully passivated alloys are suitable. This excludes families of 13%Cr, which have shown poor resistance in these applications, and potentially 22%Cr duplex stainless steel. Furthermore, handling of raw sea water requires a minimum specification of super duplex stainless steel. Additional choices for water injection applications include internal plastic-coated tubulars using a phenolic compound, glass reinforced epoxy (GRE) for low pressure water systems, GRE-lined CLAS and polyethylene-lined CLAS, which has shown promising performance for these applications, are outlined in Figure 14.2 and referred to in Chapter 9.

14.4 System Corrosivity

Dry (non-water-containing) hydrocarbons are not corrosive in the temperature range encountered in hydrocarbon production. However, the hydrocarbon phase (oil, gas, or gas condensate) is normally co-produced with water and this aqueous phase in dynamic equilibrium with co-produced acidic gases and organic acids is the cause of undesirable corrosion and damage. The main types of internal corrosion threat are briefly described here and in more detail in Chapters 1, 5, 6, 7 and 11 and should be included and considered in the overall materials optimisation strategy.

14.4.1 CO$_2$ Corrosion

As discussed in Chapter 5, several prediction models for CO$_2$ corrosion or 'sweet corrosion' of oil and gas production are available. In particular, these models have differing approaches in accounting for oil wetting, the effect of protective corrosion films, the presence of H$_2$S and other influential parameters which account for much of the differences in their respective outcomes. All the models are capable of predicting the high corrosion rates found in systems with low pH and moderate temperature, while the models can predict rather different results for situations at high temperature and high pH, where protective corrosion films may form.

An effective strategy can be successfully achieved by informed and careful choice and use of corrosion prediction modelling to come up with a judgement on the severity of the CO$_2$ corrosion threat as an input to the materials optimisation process.

14.4.2 H$_2$S Corrosion

H$_2$S corrosion or 'sour corrosion' arising from exposure to wet hydrogen sulphide has wide-ranging implications on the integrity of materials used in the industry, as outlined in Chapter 6. ISO 15156 [5] provides specific recommendations/guidelines on materials that can be deployed in H$_2$S containing environments in terms of susceptibility to cracking. Nevertheless, other types of corrosion threat in the presence of H$_2$S need to be considered carefully as the promotion of highly localised and accelerated metal-loss corrosion or localised damage can be a primary consideration. Such damage type can especially be associated with sour gas fields, although not exclusively.

14.5 Oxygen Corrosion

In water injection systems, the corrosion rate is dominated by the oxygen content and the velocity, but also affected by the chlorine content (resulting from continuous chlorination as biocide treatment; and not to be confused with chloride), residual oxygen scavenger, solids, and operating temperature as discussed in Chapter 7. The chapter contains several methods by which corrosion of CLASs as affected by operating conditions are predicted.

14.6 Metallic Materials Optimisation Methodology

The optimum choice of materials is governed by a number of key parameters including adequate mechanical properties, joining integrity, corrosion performance, weldability (where appropriate), availability, and cost. The methodology employed in the present materials optimisation strategy combines a number of these key ingredients, focusing on materials with a proven track record while describing attributes essential for such a strategy, as outlined in Figure 14.3.

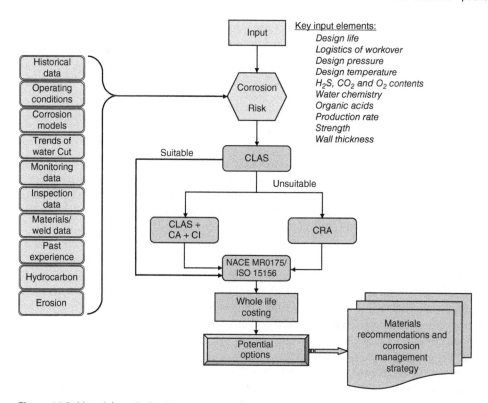

Figure 14.3 Materials optimisation strategy roadmap.

14.7 Materials Options

Several categories of alloy are used in hydrocarbon production to enable successful and trouble-free operations. Effective corrosion mitigation in hydrocarbon production has been achieved through the use of conventional grades of CLASs, inhibited CLASs, the addition of a corrosion allowance, 13%Cr, or other CRAs. Their choice is governed primarily by the prevailing system corrosivity as determined by production conditions, solution chemistry, acidic gases, and the hydrodynamic parameters; although the likelihood/frequency and severity of excursions in any one condition must also be considered and may also need to be factored in.

This section briefly describes the primary alloy systems used in hydrocarbon production with more detail given in Chapters 2 and 3.

14.7.1 Carbon and Low Alloy Steels (CLASs)

CLAS is the principal material of construction. CLAS in the majority of applications is the first, optimum, and base-case choice, subject to ensuring through effective corrosion management its resistance to the prevailing conditions.

14.7.2 Low Cr Containing Steels

A new generation of low Cr containing steels with 1–5%Cr are being manufactured and offer slightly improved CO_2 corrosion resistance. Their deployment has primarily been considered for well completion applications, particularly in low CO_2 containing conditions.

14.7.3 Families of 13%Cr Steels

Families of 13%Cr stainless steel have been used extensively for sub-surface applications in sweet and mildly sour production conditions where they exhibit good corrosion resistance. They are generally selected for their relatively high strength and satisfactory resistance to CO_2 corrosion threat. They normally contain 13%Cr steel and provide a degree of passivity in hydrocarbon production, hence its low corrosion rate.

A more recent generation of 13%Cr super martensitic grades is becoming widely available. These are used particularly for subsurface application as tubular in well completions. The alloy is the development of conventional ISO 11960/API 5CT grade 13%Cr steel to which additional Cr (13-17%Cr) and alloying elements of Ni, Mo, and Cu have been added. The super martensitic grades combine high strength and low-temperature toughness with improved corrosion resistance in sweet production conditions compared with 13%Cr.

A new generation of low C weldable 13%Cr is gaining grounds for infield flowlines and pipelines in CO_2 containing conditions. The challenge here is to address weldability, the presence of trace H_2S and compatibility with external cathodic protection (CP). Nevertheless, more recently, weldability is less of an issue provided appropriate post-weld heat treatment is undertaken.

14.7.4 Other CRAs

Other types of CRAs are covered in Chapter 3 and include several categories of alloy containing varying amounts of chromium, nickel, molybdenum, iron, and other alloying elements to offer superior corrosion performance. CRAs rely on the formation of a passive film to render them corrosion-resistant. These include duplex stainless steels, high alloyed austenitic stainless steels, nickel based alloys, and others. They are invariably more costly than CLASs – a cost penalty of >6–8%; although their selection may be essential for critical application or where other low grade alloys do not offer adequate performance. CRAs are primarily restricted to use for well completion and as internal cladding of manifolds and internal cladding of risers due to their relative cost. However, based on whole life cost comparison, they may become economical for specific applications.

14.8 Internal Corrosion Mitigation Methods

Based on the combination of elements outlined in Figure 14.1, the most appropriate materials and corrosion mitigation option need to be explored. An intermediate step and a transition bridging the economic gap between CLAS to CRA and other options shown in Figure 14.2 are briefly discussed in this section; more detail can be found in specific chapters dealing with respective applications and systems. These include deployment of a corrosion inhibitor and the introduction of a corrosion allowance (CA), the introduction of coatings and linings, all intended to extend the applicability of CLAS to harsher conditions.

14.8.1   Corrosion Inhibition (CI)

The successful use of corrosion inhibitors (covered in Chapter 8) will depend on many factors. Their deployment is application-specific and while typically used in pipelines and trunk lines, their deployment subsurface in wells is subject to logistics and operator philosophy. These are dictated by on- or offshore location, inhibitor availability, and CAPEX constrained operations.

14.8.2   Corrosion Allowance (CA)

The CA is defined as an additional wall thickness to cater for a residual level of corrosion occurring over the design life, and generally is used in conjunction with continuous treatment with a corrosion inhibitor described in detail in Chapter 8. The CA is designed above and over the wall thickness that is required for the mechanical integrity and pressure containment. By determining the potential mitigated rate of the corrosion threat and the expected service life of a unit, the additional wall thickness to cater as the CA is calculated at the design stage. Note that poor in-service management of inhibitor treatment, changing operating conditions, and a desire to extend the service life beyond that of the original design life can complicate the remaining usefulness of the in-service degraded CA. Access to quality corrosion monitoring and inspection data (described in Chapter 10) will be key to making an informed judgement call here.

Again, the CA is subject to the operator philosophy of design. While some consider CA to compensate for the eventual metal loss expected over the life of the unit, others use it as a safety consideration.

14.8.3   pH Stabilisation

pH stabilisation is yet another method by which CO_2 corrosion threat can be controlled. It has been deployed in a limited number of cases and is an option to be considered. However, it is not commonly used and invariably needs support from inhibitor treatment potentially until a line is satisfactorily covered by a layer/film with a protective corrosion product. It should be noted that this option is not as easy to satisfactorily apply and manage compared to an inhibitor treatment. The need to consider its use arises in very exceptional circumstances and where a highly experienced operations team to manage the deployment is in place [6].

pH stabilisation promotes the formation of a very protective iron carbonate corrosion film to render corrosion to an acceptable level. This is achieved by raising the in-situ pH by strong buffering with a base that will lead to a concentration of bicarbonate significantly higher than normally found in natural formation waters. This method has been successfully applied to wet gas transmission pipelines and flowlines. The method has mainly been applied in sweet systems (CO_2 only) with some limited examples when H_2S

has been present. This system is potentially viable when monoethylene glycol (MEG) is required as a hydrate preventer. A base is added to the bulk MEG/water phase that promotes the formation of very protective corrosion films. The method is only applicable to systems where no aquifer water breakthrough is occurring due to the risk of calcium carbonate scaling [6].

The pH stabilisation may be used in sour systems, but sulphide films will be formed instead of carbonate films and a higher concentration of the stabiliser may be required.

14.8.4 Internal Coatings and Linings

An alternative option to mitigate the corrosion threat to CLAS is the use of internal coatings or linings, using non-metallic options. This is dealt with above and in Chapter 9.

14.9 Whole Life Cost (WLC) Analysis

Whole life cost analysis provides the best means by which capital expenditure is optimised to offer the highest rate of return on investment. After drilling, materials is the highest cost item in hydrocarbon production and a materials optimisation strategy can lead to economies while meeting operational constraints and safety. An optimum materials strategy implies choosing the materials of construction to enable a balance of minimum CAPEX with acceptable operating expenses (OPEX) to maximise the project value. In such a strategy, the emphasis should be placed on making the correct economic selection between CLASs, inhibited CLAS, and other alternative non-metallic materials or CRAs.

The method used in assessing WLC is as follows (Eq. 14.1) [7, 8]:

$$WLC = AC + IC + \sum_{n=1}^{N} \frac{OC}{\left(1+i\right)^{n}} \tag{14.1}$$

where:

WLC = whole life cost
AC = initial capital cost
IC = installation cost
OC = operating and maintenance cost
i = discount rate
N = design life (years)
n = year of the event.

This method has been expanded to include costs associated with lost production, replacement, and residual value [7, 8]. While this OPEX-loaded option is favoured by conventional net present value (NPV) analysis; risk integrity management (IM) needs to be built into the process rather than as a retrofitted option to get a true picture, as briefly discussed in Chapter 17.

14.10   Materials Optimisation Strategy

A materials optimisation strategy requires the integration of key parameters to allow the selection of the most suitable, safe, and economical material option and corrosion control procedure. The parameters captured in such a strategy should take advantage of two key elements of trusted methods as well as innovative solutions. It should reflect past experience – successes and failures (lessons learnt) – and should use innovative means and materials to allow progressive solutions. Several parameters can be included in the overall strategy. A simplified roadmap for materials optimisation is shown in Figure 14.3.

A notable example of these parameters necessary to be taken on board is as follows:

- *Corrosion risk* should be defined by taking account of historical data, trends of water cut, inspection and monitoring data, flow dynamics, materials data, pipe inclination, and the influence of phase ratios on the onset of oil in water emulsion breakout allowing a significantly better indication of the potential risk. Risk is a combination of potential and consequence, as discussed in Chapter 16. In addition, there is a need to consider the criticality of the facility, equipment, etc., e.g. is it safety-critical?; how easy can loss of containment be confined?
- *Operating conditions* cover the most influential parameters which affect the choice of materials including design life, the logistics of workover, the temperature and pressure, strength requirements, environmental conditions, and the production rate.
- *Corrosivity assessment* for both sweet/sour production and, where applicable, water injection is built through cross-referencing of available models. In using corrosion prediction models, there is a need to exercise great care in how to cross-reference models (see Chapter 5) – at worst this can potentially become confusing or the model giving the most favourable answer is used. What becomes important here is consistency. This implies that a model could be selected having considered all the available options. Thereafter, once one is comfortable and confident in the use of a particular model, stay with it without losing the site of other models; and always looking for relevant field analogues and experience to compare and benchmark against. Furthermore, understand what level or accuracy of predicted rate should be realistically expected – at best, one decimal place on mm/year damage rate, with caution. In the use of models, the system corrosivity of the production scenarios is defined, combining corrosion prediction models, hence taking advantage of all the approaches incorporating laboratory evaluations, theoretical calculations with extensive field experience, and the influential role of organic acids. Corrosivity in injection facilities is differentiated and prediction is carried out using respective modified models.
- *Erosional velocity* is included to ensure that the operating regime does not lead to velocities beyond which erosion can become likely. Increasingly production from unconsolidated reservoirs is becoming common. This is challenging to address and places a very high dependence on installation of downhole sand screens and the strategic placement of erosion probes and sand monitoring downstream. Also erosion-corrosion affecting CLASs and 13%Cr results in a synergist effect on the resulting metal loss rate and needs careful consideration.

- *The window of application* encapsulating the domains within which alloys have proven track records by considering available industry wide data, international standards, information on proprietary grades, and reliable laboratory evidence. This is not a parameter but needs to be included as part of the overall selection process.
- *Whole life costing* to provide the best means by which capital expenditure is optimised to offer the highest rate of return on investment, taking account of costs associated with lost production, replacement, and residual value.

Throughout the process, a methodical approach to performance evaluation needs to be put in place and be implemented. This provides a flexible structure to allow realistic testing to enable the input of complementary data to provide further confidence in their application.

The simple overall approach to the optimisation strategy, outlined in Figure 14.3, captures these necessary steps in finalising materials choice. This follows a methodical route to highlighting options and the most appropriate and cost-effective materials. The strategy outperforms similar models through the unique integration of key parameters. While the majority of these parameters have been discussed extensively in the past and covered in earlier chapters, the overriding element is the combined influence of the hydrocarbons phase and the flow dynamic in which the onset of an emulsion breakout may be predicted. The onset of emulsion breakout occurs when a transition from the water-in-oil emulsion to the oil-in-water phase takes place. Furthermore, the vital influence of organic acid – notably acetic acid – in the strategy has enabled a more realistic picture of the performance in CO_2 containing environments. The strategy is applicable to the optimisation of materials for all applications including downhole completions, surface, and transportation facilities.

14.11 Summary

The approach taken to optimise materials for both production and injection systems is covered here, focusing on environmental, operational, and hydrodynamic parameters, bearing in mind the whole life costing. A simple methodology is presented based on the use of past successes and lessons learnt in effective use of CLASs and the integration of key parameters to allow the selection of the most suitable, safe, and economical material option and corrosion control measures.

Particular reference has been made to elements of production facilities with respective materials choices and corrosion mitigation strategies.

The focus in any design should be placed on WLC and use of alloys with a proven past track record.

References

1 ISO, Petroleum, petrochemical and natural gas industries – Materials selection and corrosion control for oil and gas production systems. ISO 21457, 2010.
2 Heidersbach, R. (2011). *Metallurgy and Corrosion Control in Oil and Gas Production*. Wiley.

3 MB Kermani, JC Gonzales, GL Turconi, et al., Materials optimisation in hydrocarbon production. NACE Annual Corrosion Conference, Paper No. 05111, 2005.
4 Craig, B. (2004). *Oilfield Metallurgy and Corrosion*, 3e. NACE International.
5 ISO, Petroleum and natural gas industries – materials for use in H_2S containing environments in oil and gas production. ISO 15156, 2005.
6 S Olsen, Corrosion control by inhibition, environmental aspects, and pH control: Part II: Corrosion control by pH stabilization. NACE Annual Corrosion Conference, Paper No. 06683, 2006.
7 P Jackman (ed.), A Working Party Report on the Life Cycle Costing of Corrosion in the Oil and Gas Industry (EFC 32): A Guideline. Publication No. 32, European Federation of Corrosion, 2003.
8 ISO, Petroleum and natural gas industries— life-cycle costing, Part 3: Implementation guidelines. ISO 15663-3, 2001.

Bibliography

API, Recommended practice for flexible pipe. API 17B, 2014.

API, Specification for bonded flexible pipe. API 17J, 2014.

ISO, Petroleum and natural gas industries – steel pipe used as casing or tubing for wells, ISO 11960/API 5CT, 2004.

Jones, L.W. (1988). *Corrosion and Water Technology for Petroleum Producers*. Tulsa, OK: OGCI Publications.

Kermani, B, Materials optimisation for oil and gas sour production. NACE Annual Corrosion Conference, Paper No. 00156, 2000.

Kermani B and Chevrot, T, Pipeline corrosion management: a compendium. NACE Annual Corrosion Conference, Paper No. 3723, 2014.

Kermani B and Dagurre F, Materials optimization for CO_2 transportation in CO_2 capture and storage. NACE Annual Corrosion Conference, Paper No. 10334, 2010.

Kermani, B. and Morshed, A. (2003). CO_2 corrosion in oil and gas production: a compendium. *Corrosion* 59: 659–683.

15

Non-metallic Materials: Elastomer Seals and Non-metallic Liners

Non-metallic materials are an essential element of the facilities in upstream hydrocarbon operations, being widely used in a range of functions from seals and corrosion barriers to piping and structural elements. Given their role in maintaining primary or secondary containment, the selection and use of non-metallic materials should be as rigorous as that for metallic materials within any facility.

Elastomers (or rubbers) are widely used in oilfield sealing applications. These are highly elastic, polymeric materials, typically used in compression seals and bearings in a range of downhole, subsea, topsides, and pipeline applications.

Thermoplastic liners continue to find wider applications in corrosion control and also in permitting life extension of existing, ageing facilities. The materials employed are fundamentally different in nature to elastomers, having a much smaller elastic range, and the way in which they are used is therefore somewhat different. Thermoplastic liners have an extensive track record in providing a corrosion barrier within carbon steel pipelines and downhole tubing. Similarly thermoplastic materials also find wide application in unbonded flexible pipes, being used as internal intermediate and external sheaths in these complex pipe structures. This will be discussed in Section 15.3.4.

This chapter first briefly describes elastomeric materials commonly used in hydrocarbon production and outlines drawbacks in the use of such materials when considering their selection. Section 15.2 has also been assigned to non-metallic liners, outlining key issues in their performance and range of materials used.

The chapter is not intended to be exhaustive. Rather, it describes further applications of non-metallic materials, not covered elsewhere in this publication, specifically elastomeric seals, non-metallic liners, and flexible pipes. It focuses on dealing with typical engineering questions, including types of materials, potential challenges, and means of degradation with respect to materials selection.

15.1 Elastomer Seals

The properties and capabilities of elastomeric materials are strongly affected by the specific ingredients used in their manufacture, including the base polymer, fillers, and additives. Likewise the curing system and the process of manufacturing are also important. Selection of an elastomeric seal can therefore be complex, and, as with

Corrosion and Materials in Hydrocarbon Production: A Compendium of Operational and Engineering Aspects, First Edition. Bijan Kermani and Don Harrop.
© 2019 John Wiley & Sons Ltd. This Work is a co-publication between John Wiley & Sons Ltd and ASME Press.

metallic materials, needs to be based on the required properties and performance when exposed to the expected service environments.

In such a selection process, elastomeric seals have to meet a number of requirements including: (i) chemical resistance to the environment to which they are exposed; (ii) capability to perform at the in-service temperature and pressure; and, finally, (iii) retention of adequate mechanical properties through the required lifetime.

15.1.1 Commonly Used Elastomer Materials for Upstream Hydrocarbon Service

There are a number of elastomeric materials that are commonly used in upstream operations. Typically, nitrile (NBR), hydrogenated nitrile (HNBR) and fluoroelastomer (FKM, TFEP and FFKM) materials are preferred for elastomer seals. These are commonly referred to by specific polymer tradenames, e.g. FKMs may be called 'Vitons' and TFEPs called 'Aflas'. This is not strictly correct, and there is a wider supplier base available for most options.

Within each of these families of materials, a relatively wide range of capabilities is possible, depending on the precise base polymer employed and the curing system and fillers which are used. A number of these are outlined in the following sections.

15.1.1.1 Nitrile Rubber (NBR)

NBR is the general term for a broad range of acrylonitrile butadiene copolymers. The acrylonitrile content of nitrile sealing compounds varies considerably (18–50%) and this markedly influences the physical, mechanical, and chemical properties of the material.

NBRs typically have very good mechanical properties when compared with other oilfield elastomers and they are widely used in both aliphatic hydrocarbon fluids and water-based service, in the temperature range -20°C–20°C. Below the lower temperature limit, NBRs become too brittle and stiff to function as seals, above the upper temperature limit, degradation may occur more rapidly.

Limitations for NBRs include application in aromatic hydrocarbons, strong acids, bromide brines, and H_2S.

15.1.1.2 Hydrogenated Nitrile Rubber (HNBR)

HNBR represents an improvement on standard NBRs. The hydrogenation process removes chemically susceptible 'double bonds' in the polymer chain, resulting in a higher upper temperature limit (to 150°C), superior mechanical characteristics, and somewhat better chemical resistance.

15.1.1.3 Fluorocarbon Rubber (FKM)

FKMs have excellent resistance to high temperatures (up to 200°C), and to a wide range of oilfield fluids and chemicals. However, improved chemical resistance generally comes at the expense of mechanical properties. A wide range of grades is available within this family of materials. Their main limitation, in the oilfield, relates to a susceptibility to chemical attack by amine-based corrosion inhibitors above 100°C or so.

15.1.1.4 Tetrafluoroethylene-Propylene Rubber (TFEP)

TFEPs have a similar range of temperature and chemical capabilities to the FKM materials, but the advantage of good resistance to amines. Low temperature capability, resistance to aromatic hydrocarbons, and overall mechanical properties are limited.

15.1.1.5 Perfluoroelastomers (FFKM)

FFKMs offer almost universal chemical resistance, but at the expense of loss in mechanical properties. These materials typically find widest application in high temperature (>150 °C), chemically demanding service. Fluoropolymer-encapsulated elastomer O-rings are typically not recommended in any oilfield application.

15.1.2 Key Potential Failure Modes

A number of key failure modes can affect elastomer seals. A group of very common failure modes relate to the elastomer material being used outside its working temperature range or in fluids with which it is incompatible, either chemically or physically. This can lead to chemical embrittlement (an example of which is shown in Figure 15.1), softening, compression set, large volume changes, and loss of elasticity at low temperature – any or all of which can lead to a seal failing.

In addition, two pressure-related failure modes need to be taken into account when selecting seal materials and designs: extrusion and gas decompression damage.

Extrusion damage occurs when a rubber seal is forced into the gap which it is sealing as a result of the applied pressure. As shown in Figures 15.2a, 15.2b, this failure mode begins with a 'lip' being formed as the elastomer forced into the gap is mechanically

Figure 15.1 An example of heat ageing of a rubber seal. (see colour plate section).

(a)   (b)

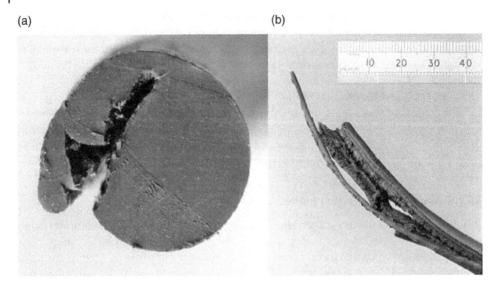

Figure 15.2 Examples of extrusion damage to rubber seals. (see colour plate section).

(a)   (b)

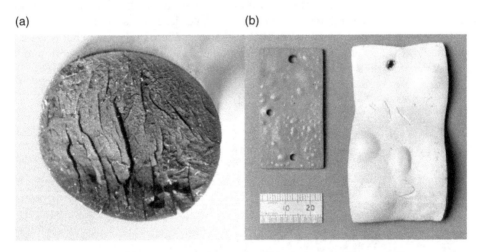

Figure 15.3 Examples of decompression damage to rubber. (see colour plate section).

torn from the seal. Extrusion is a function of many factors including materials properties, the effect of fluids on those properties, and the design of the seal housing within equipment.

Gas decompression damage occurs primarily in dry gas duty. Qualitatively at least, gas decompression damage is well understood. When a rubber seal is put under pressure, gas dissolves within the rubber. If the applied pressure is released, dissolved gas cannot move quickly enough out of the rubber to avoid supersaturation of the gas within the seal. Gas bubbles may then nucleate and grow within the rubber leading to blisters and/or internal cracks, as shown in Figures 15.3a, 15.3b, and eventually seal failure.

Gas decompression damage can potentially occur in all gas handling processes, e.g. production, compression, injection, gas lift. 'Dry' conditions, however, are generally most severe, as the presence of any liquid affects gas transport in such a way as to limit the potential severity of damage. The prefixes 'Explosive' and 'Rapid', which are often applied to decompression damage, should be regarded as misnomers, since decompression damage can occur even when pressure is let down over many hours or even days, depending on seal geometry and operating conditions.

The threshold pressure above which damage occurs is linked to the hardness of the rubber, e.g. 250 psi (1.7 MPa) for 50 Shore A material, 500 psi (3.5 MPa) for 90 Shore A.[2] Damage generally increases with pressure, but is less sensitive to temperature within the normal working range of the material.

Field experience shows that the majority of gas decompression failures can be linked to a combination of factors

- large section seals, e.g. > 10 mm;
- soft materials, e.g. 50–70 Shore A;
- non-decompression resistant materials.

Seals in dynamic service exhibit a series of additional failure modes, including wear, spiral failure, and hysteresis heat build-up.

15.1.3 Seal and Materials Selection

Elastomer seal materials are selected through consideration of the components to be sealed and the temperature, pressure, and fluids to which they will be exposed. Qualification of seals and materials performance of any component or system is typically carried out using a combination of materials and system testing, taking account of the time- and temperature-dependent properties of the materials involved. Finite element analysis (FEA) modelling has a limited capability, primarily as a result of complex, non-linear, and time dependent materials properties, but has proven vital in some applications.

Initially, seal design is selected according to the components to be sealed. Many static seals employ simple O-rings. This robust design is widely used, and straightforward to design, manufacture, and install. Larger seals can take the form of packers, of one type or another, which are similarly activated through mechanical compression. In dynamic service, T-seals may be selected or any one of a range of shaped seals designed for specific purposes.

Further selection criteria are then used to mitigate the main failure modes, discussed earlier in this chapter.

15.1.3.1 Degradation Due to Temperature and Chemical Environment

Degradation due to temperature and chemical environment can be mitigated by selecting the proper material type for any given service. Consideration is typically first made of elastomer capabilities versus the design and operating temperature range and the main fluids to be encountered. This is achieved using compatibility data available from across the supply chain, from polymer and seal suppliers to oilfield equipment manufacturers. End users typically also have a good deal of data specific to their business.

In addition, consideration also has to be given to chemicals which are added to the process either in small quantities or only occasionally. The impact of these on elastomer seals often has to be assessed based on previous field experience and technical judgement.

15.1.3.2 Extrusion

Extrusion is most easily mitigated by the incorporation of anti-extrusion devices, such as scarf cut PTFE (to 5000 psi, 24.5 MPa) or polyether-ether-ketone (PEEK) (to 15 000 psi, 103.5 MPa) back-up rings. These materials have outstanding chemical resistance to virtually all oilfield environments, and a very wide operating temperature range (<−150 °C to >250 °C). They can be used on their own in spring energised seals, but never on their own as O-rings or any other compression seal design. Material properties can be markedly affected by the method of manufacture. For example, hot melt processed PEEK has a much higher ultimate elongation than compression moulded PEEK. Also, isostatic moulding produces much better materials properties than compression moulding, e.g. with PTFE materials.

PTFE's resistance to deformation under load is substantially improved through the incorporation of fillers and reinforcements, such as graphite, molybdenum disulphide, asbestos fibres and glass fibres or flakes.

Hard rubber back-up rings are generally unacceptable.

15.1.3.3 Decompression Damage

Avoiding decompression damage may involve the use of: 'decompression-resistant' materials; small section seals (e.g. with a cross-section ≤5.33 mm); high groove fill seal designs (to stop expansion of internal cracks); and controlled decompression rates.

Decompression performance is at least partially defined by Annex F of ISO 23936-2. Seal suppliers can provide test certificates demonstrating the performance of O-ring seals under particular operating conditions. This data can be used to select particular materials for particular applications, remembering the chemical demands which may also be placed on decompression-resistant materials, e.g. methanol, high CO_2 content, amine corrosion inhibitors. End users may prefer to use specific materials which have proven capabilities in their operations. Limited predictive modelling/FEA of gas decompression damage is also now available. This can help qualify materials for a particular service, and can reduce the need for time-consuming and expensive seal testing.

It can be argued that seals are generally surrounded by large sections of steel and are unlikely to be affected by transient low temperatures. Calculations may be required to justify this, but in the majority of cases, seal selection in high pressure (HP) gas service should be made primarily for gas decompression, rather than transient low temperature capability. It should be noted that elastomer materials capable of continuous very low temperature service are generally not decompression-resistant. Extended valve stems can be useful in such applications.

Where no suitable rubber seals are available for a particular HP gas service, sprung PTFE seals should be considered.

15.1.4 Project-Specific Elastomer Seal Selection Guidelines

Any onshore or offshore oilfield development consists of a very wide range of services, and it is very rarely possible to select a single seal material for all applications. Seals must be selected for specific service. One way to record this information is in a project-specific seal selection document [1]. This records the services and gives recommendations regarding seal materials for each application. Detailed design and operational data are provided as inputs by the project.

15.2 Non-metallic Liner Options for Corrosion Control

There are a number of 'pull-through' liner technologies which can offer cost-effective solutions to mitigate internal corrosion challenges in carbon steel flowlines to some extent dealt with in Chapter 9. Rehabilitation options include the use of reinforced thermoplastic pipes (RTP) as loose-fitting slip liners and tight-fitting thermoplastic liners.

In addition, stand-alone RTPs and thermoplastic liners are now also being widely used for new build lines, as an alternative to solutions using higher metallurgies. This section briefly describes the current available technologies.

15.2.1 Reinforced Thermoplastic Pipe (RTP)

Flexible, corrosion-resistant, high pressure plastic pipe is a proven option for small diameter onshore flowline applications, having established a considerable track record in the US, Canada, and the Middle East. There is also a developing offshore track record. These products are now typically referred to as RTPs.

RTPs are designed as stand-alone pipes, but have also been widely used as slip liners inserted through the conduit of a corroded steel pipe. They are typically available for pressures up to around 2500 psi (17.2 MPa) in sizes up to 6" (150 mm) diameter. Higher pressures are possible at small diameter.

15.2.1.1 RTP Structure
RTPs consist of a continuous thermoplastic liner overwound with fibre reinforcement. The plastic liner, in contact with the transported fluid, can be manufactured from one of a range of materials commonly used in the oilfield. The majority of RTP products use high density polyethylene (HDPE) liners, giving a baseline capability of around 60 °C in water and gas, somewhat lower in hydrocarbon liquids. In water service, 80 °C may be possible with higher performance polyethylene raised temperature (PE-RT) grades. Alternative products using polyamide (PA), polyphenylene sulphide (PPS) or polyether-ether-ketone (PEEK) liners are available, offering higher temperature capability and better hydrocarbon resistance.

The reinforcement is designed to take the pressure and other mechanical loads on the pipe. Several types of reinforcement are commonly used: aramid fibres, glass fibres, and glass epoxy composites. Metal strip and cord-reinforced products have also now been made available for subsea applications. These overcome buoyancy issues, but introduce corrosion and cracking concerns.

The majority of RTP products have a thermoplastic outer cover, to protect the pipe structure.

15.2.1.2 RTP Qualification
While each RTP product is different in terms of the materials and methods used to make the pipe, there are sufficient similarities in design to enable a range of diverse products to be qualified using similar principles. Onshore use of RTPs has been covered

for the past 10 years by API RP/Spec 15S. This was initially aimed solely at RTP with non-metallic reinforcements, but now allows metallic wires and strips.

Suppliers aiming for higher-pressure, offshore applications have always been reluctant to embrace API Spec 15S due to its emphasis on long-term (10 000 hours) pipe testing. Initially some preferred API Spec 17 J, particularly for metallic reinforcement, but DNV RP F119 is now often quoted. This document enables bespoke pipe designs with limited full-scale pipe testing. It consequently has some perceived limitations, particularly with respect to end fitting design and qualification.

In all these standards, RTP manufacturers are required to demonstrate the capability of their product by undertaking a series of pipe qualification tests. These include: pressure rating using long-term (10 000 hours) testing of pipe and end fittings; characterisation of minimum bend radius for storage and transportation, and for operation; characterisation of axial load capability, to be used, for example, in installation methodologies; demonstration of capability of the product to handle gas service, and performance of the product in UV.

There is a range of engineering design issues that need to be worked through with each product. Some, such as internal surface roughness, heat transfer coefficient, and pipe expansion due to pressure and temperature are common to all. Some, such as the corrosion and cracking limits on steel reinforcements, apply only to certain products.

The industry standards rely on suppliers being able to consistently manufacture products within repeatable properties. Manufacturing quality and good manufacturing quality plans and inspection and test plans are key to a successful project using RTPs. Fortunately, well-established QA/QC procedures exist for thermoplastic and composite pipe, which can easily be adapted for high pressure plastic pipe. End fittings should be fitted at the manufacturer's premises, as far as possible, to improve quality control and speed up field installation.

15.2.1.3 RTP Installation

Installation of RTPs is typically from a reel. It is essential to limit the tensile load that is applied to the pipe during deployment, and to control the bend radius of the pipe.

Onshore, RTPs can be pulled through an existing pipe with minimal ground excavation. Usually the host pipe ID is much larger than the inserted RTP, e.g. 10 inches host for a 4 inches RTP. This difference decreases pulling loads significantly. The host pipe must be accessible at both ends, and have no obstructions to the passage of the RTP, e.g. bends with radii smaller than 1.5x the minimum bend radius of the RTP, or internal weld protrusions.

The installation process involves correcting any deficiencies in the host pipe, passing a pulling cable through the line, attaching the cable to the RTP and pulling the RTP back to the winch. Finally, the RTP is terminated with appropriate end fittings.

Offshore pull-through installations are similar, with tight bends, e.g. at riser transitions, typically problematic. Some RTP suppliers recommend that subsea pull-throughs be done with a flooded pipe, which reduces pulling loads significantly. The track record for offshore rehabilitation is limited at the moment, but will continue to grow as more operators embrace this technology.

For stand-alone applications, RTPs may be simply laid on the ground. Key considerations include the potential for UV damage and the need to anchor or restrain the pipe during both hydrotest and service. Trenching is also often employed, with trench depth and backfill selected to avoid external damage to the pipe. Ploughing methods and directional boring have also been used in RTP installations.

15.2.2 Thermoplastic Liners

Tight fitting thermoplastic liners can be inserted into carbon steel flowlines using a number of proprietary technologies. Such liners use the steel pipe for pressure containment, with the thermoplastic acting as a corrosion barrier.

This technology is well proven onshore for water and hydrocarbon liquid service, being widely used in the US, Canada, and the Middle East. Use in high gas oil ratio (GOR) production service has been more problematic, due to the poor collapse resistance of plain PE liners. This has been managed operationally, or by liner replacement upon failure. Technology incorporating external grooves in the liner, to enable more efficient evacuation of permeated gases, has been qualified and has seen limited use in the field.

Offshore, plastic-lined water injection flowlines and riser towers are now standard practice for handling sea water and commingled sea water/produced water. Tight fit PE liners are a very effective way of avoiding the bottom of line corrosion prevalent in carbon steel, subsea WI lines.

15.2.2.1 Plastic Liner Installation

Tight liners are inserted by reducing the outside diameter of the plastic pipe to something less than the internal diameter of the steel host. There are two leading techniques: roller box reduction and conical die reduction. Both have been widely used onshore, and in offshore projects executed with lining done onshore. An example of roller box reduction equipment is shown in Figure 15.4. The overall installation procedure is similar for both reduction methods and both achieve the same end, but with different stress conditions on the liner during installation. As long as tension remains on the liner, during installation, the diameter will remain reduced. When tension is removed, the liner reverts to a larger diameter (and shorter length) to become tightly fitted in the host pipe.

The technology is typically limited by the bends in the pipeline and the pipeline topography. New lines can be designed to accommodate the liner by using large radius swept bends of at least 20x (and preferably 50x) nominal pipeline diameter. The access points for liner installation can be planned and left uncovered after the rest of the pipeline is buried to facilitate access for liner installation.

In onshore rehabilitation projects, before liner insertion, the host pipeline must be prepared to receive the liner. This usually involves excavating the host pipe at convenient access points, cutting the pipe and attaching a flanged end fitting suitable for use as a liner termination to create a pipeline segment to be lined. Elbows and short radius bends must be cut out and replaced with swept bends.

Offshore use has involved the reeled or towed installation of MDPE/HDPE-lined carbon steel, with lining of flowline and riser stalks carried out onshore before reeling. Several different proprietary liner connections have been used to terminate the liner stalks. Some connections have now also been proven in fatigue testing, with connection performance driven by the steel weld class rather than the polymer termination details. This clears the way for the use of PE-lined pipe in steel catenary riser (SCR) and other riser applications.

Figure 15.4 An example of roller box reduction equipment. (see colour plate section).

To date, there have been no examples of offshore rehabilitation using tight fit liners, although design and feasibility studies have been completed.

15.2.3 Plastic Liner Materials

In all cases, HDPE options are typically capable of ~60 °C in water and ~50 °C in hydrocarbons. Newer polyethylene of raised temperature (PE-RT) resins may extend this to 80 °C in water. Other materials options exist for higher temperatures, e.g. temperatures up to 90 °C are possible with nylon (PA-11) in hydrocarbon production.

Plastic liners are typically manufactured from established pipeline grades, i.e. PE80 or PE100 as defined by ISO 12201-1. They have sufficient thickness, with appropriate safety factors, to resist loads induced during insertion, installation, and operation, including collapse by the action of applied vacuum and permeated gas.

15.2.3.1 Plastic Liner Qualification

Liner technology qualification should include: characterisation of a range of relevant materials properties; finite element (FE) modelling of the liner within the host pipe, including design features, compression rings with end connectors, vent holes and steel weld beads; tests of short-lined spools to verify liner collapse resistance, and elevated temperature system and connector performance; system tests of longer-lined spools under flowing conditions.

15.2.4 Summary of Technology Capabilities

The current capabilities of RTP slip liner and tight fit plastic liners are summarised in Table 15.1.

15.2.5 Other Technologies

Several other liner technologies are worth noting: (i) 'hybrid' solutions; and (ii) composite-lined downhole tubing.

15.2.5.1 'Hybrid' Solutions

Several suppliers offer a slightly different rehabilitation technology with thin RTPs which are inserted in a collapsed or folded state. Once in place within the host pipe, the RTP 'sock' is inflated to stand directly against the host pipe. This has the advantage of

Table 15.1 Typical current industry capability of non-metallic liners.[a]

Application	Options	Current status	Typical limitations
Onshore – water Injection (WI) (New build and rehab)	RTP slip liner	Field proven	2500 psi, 6″, PE < 60 °C
	Tight fit liner	Field proven	PE < 60 °C
Onshore - production (New build and Rehab)	Tight fit liner	Field proven for hydrocarbon liquids	Gas: P < 200 psi No gas: PE < 50 °C; PA-11 < 90 °C
	Grooved liner	Field proven	PE < 60 °C, 3000 psi PA-11 < 90 °C, 6000 psi
	RTP slip liner	Field proven	2500 psi, 6″ PE < 50 °C; PPS < 120 °C
Offshore – WI (New build)	Tight fit liner	Field proven	PE < 60 °C
Offshore – production (New build)	Tight fit liner	Field proven for hydrocarbon liquids	Gas: P < 200 psi No gas: PE < 50 °C; PA-11 < 90 °C
	Grooved liner	Concept proven Reeled only	PE < 60 °C, 3000 psi PA-11 < 90 °C, 6000 psi
Offshore – WI (Rehab)	Tight fit liner	Concept proven	PE < 60 °C
	RTP slip liner	Field proven	2500 psi, 6″, PE < 60 °C
Offshore – production (Rehab)	Tight fit liner	Concept proven for hydrocarbon liquids	Gas: P < 200 psi No gas: PE < 50 °C; PA-11 < 90 °C
	Grooved liner	Concept proven	PE < 60 °C, 3000 psi PA-11 < 90 °C, 6000 psi
	RTP slip liner	Field proven	2500 psi, 4″ PE < 50 °C; PPS < 120 °C

[a] *Note:* This is an updated version of a table contained in EFC Publication No 64 (Annex B).

very low pull-in loads, and hence long pull lengths, with a product which has at least some pressure holding, and hence hole-bridging, capability. This technology has a growing track record in a range of applications.

15.2.5.2 Composite-Lined Downhole Tubing

A number of suppliers offer tubing lined with glass reinforced, epoxy composite liners, per API RP 15CLT. This is further described in Chapters 9 and 14. The annulus between the liner and the host pipe is typically filled with a special type of cement, to transfer mechanical and pressure loads. Modified connections allow the liner to be properly terminated, with thermoplastic corrosion barriers used to provide continuity of corrosion performance.

Composite-lined downhole tubing has a long track record of successful onshore use in a range of corrosive service and in offshore water injection. Some liners are capable of continuous service at up to 80 °C in water-based applications.

15.3 Flexible Pipes

Flexible pipes are deployed for many applications in oil and gas production systems. Their flexibility often enables faster or more convenient installation and hook-up and provides excellent fatigue resistance in a range of harsh environments.

For offshore production and injection systems, subsea risers and jumpers are often constructed using unbonded flexible pipes. These pipes are typically engineered on a bespoke basis through API Spec 17J and RP 17B. A typical pipe structure is shown in Figure 15.5.

In considering materials selection in a flexible pipe, it is useful to take a layer-by-layer approach. While the carcass and primary pressure sheath are flow-wetted, many of the layers in a flexible pipe operate within the so-called annulus of the pipe. The

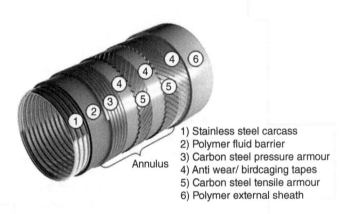

1) Stainless steel carcass
2) Polymer fluid barrier
3) Carbon steel pressure armour
4) Anti wear/ birdcaging tapes
5) Carbon steel tensile armour
6) Polymer external sheath

Annulus

Figure 15.5 Schematic drawing of an unbonded flexible pipe. *Source:* Reproduced with kind permission from TechnipFMC, France. (see colour plate section).

environment therein normally depends on permeation from the bore and external environments.

Materials selection issues have also arisen in some of the ancillary equipment associated with flexible pipes. For example, fixing and latching mechanisms have been subject to corrosion failures, particularly when not properly protected by cathodic protection (CP). This has led to a number of bend stiffeners becoming loose and sliding away from their proper location. Fatigue damage to the polyurethane bend stiffeners themselves has also been seen.

With such complicated pipe structures and so many potential threats, in-situ inspection has proven challenging, beyond visual inspection by remote operating vehicle (ROV). Some technologies are now routinely used to monitor the condition of flexibles in operation, but these are typically indirect measures of pipe condition. Polymer coupons have been widely used to assess the state of plastic sheaths, particularly nylon materials in high temperature operation. Interrogation of the annulus of flexible risers has also been widely undertaken, using volume and pressure checks to judge whether there has been sea water ingress. There has also been some use of X-ray inspection to penetrate both pipe structures and end fittings.

15.3.1 The Carcass

Flexible pipes frequently have an inner metallic carcass to provide the pipe with collapse resistance. This carcass is manufactured from thin metallic strip (sometimes multi-layered) fabricated into an interlocked (roll-formed) or corrugated tube. The material selected for the inner carcass needs to have corrosion resistance to the fluids likely to be transported in the flexible pipe, sufficient erosion resistance, and mechanical strength for collapse resistance. Typically the materials used for the inner carcass are corrosion-resistant alloys, including AISI 304L stainless steel, AISI 316L stainless steel, and various grades of duplex stainless steel.

15.3.2 Polymer Pressure Barrier/Sheath

The polymer pressure barrier/sheath contains the process fluid within the pipe. Performance criteria include collapse resistance, creep, gas permeation, chemical compatibility, and response to gas decompression. Several thermoplastic materials are widely used as pressure sheaths. Nylon materials are widely used in production service up to 60–90 °C, with PVDF used above this temperature range to around 130 °C. In water injection, polyethylene is normally used.

15.3.3 Reinforcement Wires

Layers of steel reinforcement wires give the flexible pipe its tensile and hoop (pressure-retaining) strength. Pressure armour wires provide the pressure-containing capacity (hoop strain). Normally they comprise a single layer against the inner polymer pressure sheath and often are constructed from just two continuous, shaped wires. The two shaped wires interlock with each other to prevent them from separating during service. For high-pressure pipes, an additional flat spiral wire may be applied on top of the shaped pressure armour wire to increase the hoop strain capacity of the structure.

Tensile armour wires provide the flexible pipe with its tensile/bending strength and contain the end-cap loads. They are normally plain 'rectangular section' wires. There are normally between two and four layers of wires, each containing many strands of wire, e.g. 50–60 wires/layer for a typical 10 in. ID pipe. Alternate layers are counterwound to torsionally balance the pipe. All these wires are made from carbon/low alloy steel strengthened by cold working or heat treatment (e.g. quenching and tempering). The high strength requirement means that the wires are not normally 'sour-resistant', and they can suffer from both sulphide stress cracking (SSC) or hydrogen-induced cracking (HIC).

Reinforcing wires are located in the so-called annulus between the inner and outer plastic sheaths, and the environment therein is determined by permeation of species, such as water, CO_2, H_2S, and O_2, through those plastic sheaths. The diffused species can result in a sour, low pH aqueous environment in the annulus. While diffusion rates are typically low, the way in which the annulus is or is not vented can be significant in determining the environment in which the wires operate over the life of the flexible riser. For example, flexible risers typically operate with vented annuli, while subsea jumpers do not.

So, in addition to the loss of a cross-section as a result of straightforward corrosion mechanisms, the wires can also be subject to the threat of SSC and HIC. In dynamic service, corrosion fatigue must also be assessed. This is carried out conservatively typically using SN (alternating stress v. number of cycles to failure) curves based on small-scale (single-wire) fatigue tests in simulated annulus environments.

15.3.4 External Plastic Sheath

The external plastic sheath contains the whole pipe structure. Typically this sheath is made of polyethylene for static pipe and nylon for pipe used in dynamic service. While the sheath is supposed to keep sea water out of the pipe structure, damage of the outer sheath during installation and handling has been a major operational issue. This damage has led to pipe annuli being partially flooded by sea water, which can have a huge effect on the residual strength and fatigue life of the tensile armour wires. In the worst case, annulus flooding can reduce the essentially 'infinite' fatigue life of an undamaged dynamic riser to something closer to a few years. Immediate intervention has typically been required to recover a useful fatigue life, for example, clamping off the damaged areas and implementing a continuous flushing procedure with an inhibited fluid to displace all the sea water. Even with such action, riser replacements have been required, although there are some indications that the fatigue assessment process may currently be over-conservative in this particular operating scenario [2].

Table 15.2 gives a brief summary of some of the materials selection issues involved in designing and operating flexible pipe systems.

There are several additional tape layers within a typical pipe structure. Intermediate polymer sheaths may be used in smooth bore pipes. Thermoplastic and fibre-reinforced tapes are used as anti-wear layers, between layers or metal wires in dynamic pipe, and to prevent 'bird-caging' when the pipe is put under compressive axial load. Thermal insulation may be added on top of the outer sheath where required.

Table 15.2 Material selection in flexible pipes.

Component	Options	Advantages	Disadvantages
Carcass	SS A304/A316	Good general corrosion resistance	Limited sea water corrosion resistance
	Duplex SS	Excellent general and localised corrosion resistance	More rigorous control of welding required
Internal polymer sheath	Polyethylene	Good sea water resistance	Limited hydrocarbon resistance Limited to 60 °C or so
	Nylon	Good hydrocarbon resistance	Limited resistance to acidic conditions
	PVDF	Good hydrocarbon resistance Higher temperature capability (to 130 °C)	Pipe design and operation must accommodate material sensitivity to notches and strain rates under certain conditions
Pressure amour	Carbon/low alloy steel	High strength	SSC/HIC Corrosion fatigue
External polymer sheath	Polyethylene	Good sea water resistance	Poor resistance to impact damage potentially leading to flooded pipe annulus
	Nylon	Improved dynamic performance	

15.4 Summary

The use of non-metallic components is an integral part of the materials selection challenge in hydrocarbon production. Given their frequent role in maintaining a primary or secondary containment, selection and use of these materials should be as carefully scrutinised as the metallic components within any facility. While the topic of non-metallic materials requires a more extensive overview than is permitted here, the chapter has attempted to offer an insight into the types and properties of some of the available options.

Notes

1 Hardness of non-metallic materials is normally measured using simple indentation techniques. Elastomer hardness is measured on the Shore A scale, typically ranging from 50 to 90. Thermoplastics are measured on the harder Shore D scale. Both are defined in ISO 7619-1.
2 Unbounded means that there is no direct physical or chemical bonding between the many layers that make up the construction of the flexible pipe.

References

1 S Groves, Project guidelines for selecting seals for high pressure gas duty and other oilfield service. 17th International Conference on Fluid Sealing, York, United Kingdom, 8-10 April, 2003.
2 D Charlesworth, B D'All, C Zimmerlin, et al., Operational experience of the fatigue performance of a flexible riser with a flooded annulus. Offshore Technology Conference, Paper No. OTC 22398, 2011.

Bibliography

API, Recommended practice for flexible pipe. API 17B, 2014.
API, Recommended practice for composite lined steel tubular goods. API RP 15CLT, 2007.
API, Spoolable reinforced plastic line pipe. API Spec 15S, 2016.
API, Specification for unbonded flexible pipe. API Spec 17J, 2017.
ASTM, Standard classification system for rubber products in automotive applications. ASTM D2000-12, 2017.
DNVGL, Thermoplastic composite pipes. DNVGL-RP-F119, 2015.
ISO, Elastomeric seals – Material requirements for seals used in pipes and fittings carrying gaseous fuels and hydrocarbon fluids. ISO 16010, 2005.
ISO, Petroleum, petrochemical and natural gas industries – Non-metallic materials in contact with media related to oil and gas production. ISO 23936, 2009.
ISO, Rubber, vulcanized or thermoplastic – Determination of indentation hardness – Part 1: Durometer method (shore hardness). ISO 7619-1, 2010.
Kermani, B. and Chevrot, T. (eds.) (2012). *Recommended Practice for Corrosion Management of Pipelines in Oil and Gas Production and Transportation*. Publication No. 64. Institute of Materials, European Federation of Corrosion.

16

Cathodic Protection (CP)

Where carbon and low alloy steel (CLAS) structures are in intimate contact with environmental waters or the ground, it is customary for them to be protected from external corrosion threats at the points of contact by cathodic protection (CP), often in conjunction with a protective coating system. In a similar manner, vessels or tanks used to contain various environments may be protected from internal corrosion by a combination of coatings and CP. This involves making the metal to be protected act as a cathode against an external body which is intentionally made to become anode and corrode in preference to the intended structure.

The method of CP can be by one of two principal forms: either by sacrificial means or by applying an external impressed current using a direct current (DC) source, normally a transformer rectifier. With sacrificial anode CP, anodes of a material less noble (more reactive) than iron, such as an aluminium alloy, a magnesium alloy, or zinc are attached to the steel. The anode material is cast onto a steel core which protrudes from the anode at selected locations. Each anode is attached to the structure via the ends of the exposed core usually by welding. Steel doubler pads are often provided on the structure to minimise localised stresses at the attachment points. Sacrificial anodes corrode preferentially to the steel and are consumed while the steel remains free from corrosion.

With impressed current systems, the cathodic current is applied to the steel structure via a number of 'inert' anodes from one or more direct current sources, usually transformer rectifiers. By comparison with sacrificial systems, a fewer number of impressed current anodes are required due to their much higher current outputs. While impressed current anodes are not consumed in the same way as sacrificial anodes, they nevertheless do have a finite life, which must be addressed in their selection, design, and operation.

This chapter is intended to focus on the practical aspects of CP by outlining the key parameters in the choice of impressed current over sacrificial anode protection and vice versa, specific applications in deepwater, and issues with regard to alternating current (AC) interference. The chapter is by no means exhaustive and intends only to underline practical considerations pointing out necessary references where further information can be sought.

It should be noted that the design, installation, operation, monitoring, and maintenance of CP systems constitute a highly specialised field of engineering and should only be carried out by suitably qualified and experienced personnel.

Corrosion and Materials in Hydrocarbon Production: A Compendium of Operational and Engineering Aspects, First Edition. Bijan Kermani and Don Harrop.
© 2019 John Wiley & Sons Ltd. This Work is a co-publication between John Wiley & Sons Ltd and ASME Press.

16.1   Key Points of Effectiveness

Effective corrosion prevention by CP depends upon the following:

- An accurate calculation of the total surface area to be protected, including any construction aids which are to remain after construction. While the latter may be redundant and do not require protection from corrosion, they will result in a drain on the CP current to the detriment of the structure if not accounted for.
- Details of the operating conditions during the lifetime of the structure. While these have a significant influence upon the calculation of the current requirements, they also influence the selection of the specific anode materials and the mechanical integrity of the attachment details for the various components of the CP system, i.e. the anodes, cables, etc.
- Identification of any likely electrical and physical interference effects either from, or on neighbouring, or adjoining structures and the incorporation into the design of any mitigation measures required to manage these effects
- A detailed inspection and maintenance programme throughout the life of the structure being protected.

16.2   Cathodic Protection in Environmental Waters

The implementation of CP on structures exposed to water is described in this section. Structures located in environmental waters include offshore oil and gas platforms and drilling rigs, offshore pipelines, and coastal and offshore jetties.

16.2.1   Design

The principal objective of any CP system is to achieve an effective protective potential as uniformly and economically as possible over the entire structure being protected for the full projected lifespan. Table 16.1 summarises CP potential design limits commonly in use on steel structures located in sea water.

The negative limits given in Table 16.1 reflect the need to ensure that the application of CP is not detrimental to the structure. For example, high strength steels can suffer hydrogen embrittlement as a consequence of the high hydrogen levels produced at the

Table 16.1 Protective potential limits[a].

Environment	Least negative protection potential	Negative potential limit
Aerated sea water	−0.800 Volts v. Ag/AgCl/sea water	−1.05 Volts v. Ag/AgCl/sea water[b]
Anaerobic sea water (e.g. sea bed mud)	−0.900 Volts v. Ag/AgCl/sea water	−1.05 Volts v. Ag/AgCl/sea water

[a] The corrosion potential of steel in aerated seawater, that is before the application of CP is typically −0.600 Volts v. Ag/AgCl/sea water.
[b] Ag/AgCl/sea water is the reference electrode most commonly used to establish the level of protection being achieved in chloride containing waters.

steel surface by cathodic over-protection. This is described more fully in Section 16.3. In addition, where the structure has been coated, the coating may suffer damage in the form of blistering or cathodic disbondment at extreme negative potentials.

Where CP is applied to a structure which has a protective coating, the current demand will be reduced compared to that required by an uncoated structure. In this case, coating breakdown factors have to be included to account for the presence of the coating and its progressive deterioration with time [1]. This is particularly relevant in the case of pipelines, which are always coated and cathodically protected.

One advantage when using protective coatings in conjunction with CP occurs at the end of the design life of the structure. It is not uncommon for some structures still to be required to be operational beyond their original design life and additional CP measures to be implemented accordingly. That is, the retrofitting of additional sacrificial anodes or additional impressed current facilities. Where there is a protective coating present on the steel, loss of protection of the structure is likely to occur much slower than for a bare structure, allowing more time to implement these measures.

16.2.1.1 Typical Design Considerations

The design of a CP system requires a full understanding of the geometry and dimensions of the structure, details of the anticipated operating conditions throughout its life and an accurate determination of the surface areas to be protected. Using the current density guidelines provided in national and international standards, the total current demand required for effective CP for the life of an offshore structure can be determined from the total surface area and the current density requirements [1, 2].

Conventional CP designs for offshore structures address the current density requirements at three main stages of a structures life as follows:

- *Initial current density:* the current density required to rapidly polarise the structure and encourage the formation on the steel surface of protective calcareous deposits in sea water, which in turn result in a fall in the current density required for full protection of the structure.
- *Mean current density:* the 'steady state' current density required to protect the structure throughout the vast majority of the life of the structure following the establishment of stable calcareous deposits.
- *Final current density:* the current density that may be required at the end of life to re-polarise the structure and to re-establish the calcareous deposits should they be damaged or removed during storm conditions.

Further details of recommended current density values to be used to cover each of these three stages for (i) different geographical locations around the world; and (ii) increasing water depths are described elsewhere [1, 2].

The CP design must be capable of satisfying each of the current demands established from this approach. For deepwater structures (>300 m water depths), the detailed design will require the structure to be subdivided into several sections to accommodate the change in current density requirement for protection with water depth [1].

Individual anode current output is determined by anode electrical resistance, which is in turn determined by anode geometry and the resistivity of the environment in immediate contact with it. Anode shape is a crucial factor in determining the anode electrical resistance. Long slender anodes, where the length is 5–10 times (say) the anode diameter or width, provide the lowest resistance and hence the highest current output. For long slender anodes, the anode resistance R in ohms is given by Eq. (16.1) as follows:

$$R = \frac{2.3\rho}{2\pi L}\left(\log_{10}\frac{4L}{r} - 1\right) \tag{16.1}$$

where:

ρ = the resistivity of the environment (e.g. sea water) in which the anodes are to function in ohm.cm;

L = the anode length in cm;

r = the radius of the anode in cm.

It should be noted that the anode resistance calculations are made using the anticipated values of L and r; at initial, mean, and final points in anode life. This determines the current output; initial, mean, and final and thus how many anodes of a particular size will be needed to meet the structure's current demands (also initial, mean, and final). Using simple software, it is then easy to determine the optimum anode shape to meet all three conditions most economically.

As sacrificial anodes are invariably cast with either a square or trapezoidal cross-section rather than as circular rods, an equivalent radius, r has to be derived from Eq. (16.2) and substituted for r in Eq. (16.1).

$$r = \frac{\sqrt{A}}{\Pi} \tag{16.2}$$

where: A is the cross-sectional area of the anode.

Further equations exist for the electrical resistance of alternative geometrical anode shapes, for example, pipeline bracelet anodes and flat plate anodes, which enable the current output from these alternative designs to be similarly estimated [3].

Once the electrical resistance, R of the anode has been determined, the current output of the anode, I can be derived from Ohm's law shown by Eq. (16.3) as follows:

$$I = \frac{E}{R} \tag{16.3}$$

where E is the anode to cathode closed circuit potential difference.

For sacrificial anode CP, there is also the question of anode utilisation to consider. This is the maximum portion of the anode volume, which can be used in providing CP current, before the anode ceases to provide the required current. Utilisation factors for long slender anodes of 0.9–0.95 can be achieved by attention to detail with regard to the following:

- the positioning of the steel anode core within the body of the sacrificial anode material;
- the length-to-width ratio of the anode (see above);
- the anode to structure stand-off.

Where adequate stand-off distances are not readily achievable, these can be offset to some extent by painting the inner face of the anode prior to anode installation.

The total weight, W, of sacrificial anode material required to protect a bare structure throughout its design life is given by Eq. (16.4) as follows:

$$W = \frac{\text{Mean curren Density}\left(\dfrac{\text{Amp}}{\text{m}^2}\right) \times \text{Total Surfcae Area}\left(\text{m}^2\right) \times \text{Design Life}\left(\text{hrs}\right)}{\text{Anode Material Capacity}\left(\text{Amp hrs/kg}\right) \times \text{Anode Utilisation Factor}} \quad (16.4)$$

Coating breakdown factors need to be included in Eq. (16.4) when a coating is present on the steel surface [1].

16.2.2 Sacrificial Anode Materials

Sacrificial anode CP systems installed on offshore or coastal structures are invariably based upon a distribution of anodes across the structure. The anodes are manufactured from either almost pure zinc [4], or aluminium alloys containing around 5% zinc and 0.015% of indium. With aluminium alloys, small amounts of indium are required to ensure the anodes do not passivate, or undergo non-uniform consumption. Either occurrence will prevent the anodes from providing the current required for effective protection.

The choice between zinc and aluminium alloy anodes is usually determined by either economic or overall weight considerations. Where the latter is a major consideration, aluminium anodes are usually selected because of their lower density than zinc and higher electrochemical capacity (Amp-hrs kg^{-1}). Aluminium anodes also have a more electronegative open circuit potential, which results in a marginally higher current output compared to zinc for identical anode geometry.

Magnesium alloy anodes for use in sea water usually contain aluminium (c. 5%), zinc (c. 3%), and small amounts of manganese (0.2–0.7%). However, while magnesium anodes may appear attractive from their much higher (more negative) open circuit potential compared to aluminium or zinc, they have comparatively low utilisation factors (0.55) compared to aluminium alloys and zinc (see above). In addition, the high magnesium anode driving voltage may be detrimental in the case of high strength steels and can also result in damage to protective coatings where they are present in the immediate vicinity. For these reasons, magnesium alloy anodes are not normally used to protect offshore or coastal structures, but do find application in soils, and fresh or brackish waters, where their higher driving voltage is required to overcome the higher electrical resistance of the environment compared to open sea water.

Table 16.2 contains a broad comparison of the relevant electrochemical properties of the three anode materials described above. A more detailed description of

Table 16.2 Typical sacrificial anode material properties.

Material	Open circuit potential (v. Ag/AgCl/sea water)	Electrochemical capacity in seawater (A-hr Kg^{-1})
Aluminium-zinc-indium	−1.08	2420
Zinc	−1.05	780
Magnesium-aluminium-zinc	−1.50	1230

sacrificial anode alloy compositions and the effect of alloying additions and impurities on performance can be found in the literature [3].

A detailed inspection and testing programme should be carried out during anode manufacture to ensure compliance with the detailed CP design as follows [1, 5]:

- net individual sacrificial anode material weight;
- anode dimensions, including the position of the steel core;
- adhesion of the sacrificial anode material to the steel core;
- sacrificial anode material composition;
- alloy electrochemical properties (see Table 16.1).

16.2.3 Impressed Current CP in Environmental Waters

A general reference to the case of impressed current CP systems for structures exposed to water is discussed briefly in this section.

16.2.3.1 General Considerations

Impressed current CP systems have much higher operating voltages than sacrificial anode systems and therefore far fewer impressed current anodes are required to protect the same area of steel. Typically, large sacrificial anodes of an appropriate design may each provide up to 5 Amps, while individual impressed current anodes are capable of outputs up to 100 Amps. The impressed current is provided to the structure via the anodes by an external source of DC power, usually a transformer rectifier. The current to each individual anode is controlled either by manual or automatic means. In the latter case, this will involve permanent reference electrodes located at strategic points on the structure that are either hard-wired or acoustically linked to the DC power source. Impressed current CP systems require heavy duty cabling between the anodes and the DC power source. Particular attention has to be paid to anode to cable connections, and methods of attaching cables and impressed current anodes to the structure, to ensure that these attachments are sufficiently robust to withstand the forces that will act upon them during service. In the past, a number of offshore impressed current systems have failed due to these items having inadequate strength.

With impressed current CP systems relying on far fewer higher operating voltage anodes than their sacrificial anode counterparts, anode location is critical if the minimum protective potential is to be achieved without some areas of the structure seeing excessively negative potentials (over-protection) and others only minimal protection. As there is a greater risk of cathodic disbondment or hydrogen embrittlement with impressed current systems, special measures are required to ensure such problems do not occur. With impressed current anodes mounted directly onto the structure, it is necessary to provide each anode with a surrounding dielectric shield to prevent over-protection of the steel in the immediate vicinity of the anode. Alternative arrangements include increasing the distance between the active anode element and the structure by employing cantilevered anodes, or by mounting the anodes remotely on sleds located around the structure. In the latter case, this is only practical with comparatively small structures in shallow waters.

In order to optimise impressed current anode locations, computerised mathematical modelling techniques have been found to be valuable in recent years [6].

16.2.3.2 Impressed Current Anode Materials

Numerous impressed current anode materials are available, covering a range of environments in which CP can be used effectively. Of these, platinised titanium and platinised niobium have traditionally proved to be the materials most commonly used as impressed current anodes in offshore applications. However, over the last 20–30 years mixed metal oxide anodes, comprising oxides of either ruthenium or iridium combined with tantalum or titanium oxide formed on titanium substrates, have become increasingly popular. All of these anode materials are capable of operating at current densities of up to 500 Amps m^{-2} of anode surface. Unfortunately, all impressed current anodes are consumed to a minor extent during operation and the manufacturer's recommendations with regard to applied voltage and current density limits should be followed if the required design life is to be achieved. During the design of the anodes and their method of attachment to a long-term structure, some consideration should be given to the ease of anode replacement as this may prove necessary, either before the end of the design life of the structure, or should the structure be required beyond its original design life.

16.3 Cathodic Protection and Hydrogen-Induced Cracking (HAC)

When CP is applied to a marine structure, hydrogen ions (H$^+$) in the sea water are electrochemically reduced to atomic hydrogen (H) at the surface of the structure (as discussed in Chapter 6). This is particularly a challenge when using impressed current systems where over-protection may occur.

The hydrogen atoms combine to produce hydrogen gas, which is then liberated from the surface. Unfortunately, the latter stage in this process, that is the combination of the hydrogen atoms to form hydrogen gas, is a comparatively slow process and hydrogen atoms, being very small, are able to diffuse into the metallic substrate. Depending upon the strength and metallurgical condition of the metallic substrate, these hydrogen atoms may result in embrittlement of the metallic structure and premature or catastrophic failure. The relationship between the electrical potential applied to the metallic substrate and the rate of hydrogen production at the surface of the substrate is logarithmic, that is, for every unit of potential change in the negative direction, there is a 10-fold increase in the rate of hydrogen production. Where CP is applied at potentials more negative than −800 mV v Ag/AgCl/sea water to a structure which is highly stressed and of increased susceptibility due to being of high strength, high hardness, or having a susceptible microstructure, then premature failure due to HAC (hydrogen-assisted cracking) is a risk which needs to be mitigated.

Due to their comparatively simple structural design compared to conventional towers, jack-up platforms have required the use of high strength steels (yield strengths of typically 450–700 MPa). At least one operator has installed diode-controlled sacrificial anode cathodic protection on a jack-up production platform to limit over-protection, although this may not be a totally viable or the sole solution and needs to be addressed on case-by-case basis.

Particular care is required where CP is applied to corrosion-resistant alloys (CRAs) such as 13%Cr and duplex stainless steels, and nickel-based alloys. In particular, careful control of welding and weld overlay parameters is necessary to avoid the production of localised microstructures which are susceptible to HAC [7].

16.3.1 Monitoring and Inspection

With all CP systems a detailed monitoring and inspection programme should be prepared at construction and adhered to throughout the lifetime of the structure. This should include, but not be necessarily limited to:

1) Visual inspection of all components on a periodic basis, e.g. anodes, cables, attachment details, etc. for evidence of mechanical damage, malfunction, etc.
2) Measurement of the electrical potential of the structure using a standard reference electrode (usually Ag/AgCl/sea water) at representative locations at a regular frequency to ensure a satisfactory level of CP is being achieved over the entire structure. The installation at construction of permanent reference electrodes (Ag/AgCl/sea water or zinc/sea water), either hard-wired back to a suitably accessible test point, or which can be interrogated acoustically, can prove extremely beneficial, particularly in the case of impressed current systems.
3) For impressed current systems, the frequent measurement of applied voltages and currents.

16.4 Cathodic Protection of Structures in Contact with the Ground

The case of CP systems for structures in contact with the ground is discussed briefly in this section.

16.4.1 Cathodic Protection Criteria

Structures totally or partially in contact with the ground and protected from external corrosion by CP include: buried pipelines [8], below-ground storage tanks [9, 10], above-ground storage tanks [11], buried piping networks at liquid storage or processing sites and near surface well casings [12].

Due to the low driving voltages of sacrificial anodes and high soil resistivities compared to saline water, sacrificial anode CP is only feasible on comparatively short lengths of small diameter piping; or where only small areas of a structure are in contact with the ground. That is, where the current demand for efficient protection is very low and the use of sacrificial anodes with their low current outputs is practical. In these cases only magnesium alloy anodes (see Table 16.1) have a sufficiently high driving voltage to be of practical use. With structures in contact with the ground, CP is predominantly used in conjunction with protective coatings to maximise the level and spread of the CP.

The following criteria apply to the effective CP of steel in contact with the ground:

1) A negative potential of −0.850 V with respect to a copper/saturated copper sulphate reference electrode,[1] −0.950 V if the ground is contaminated with sulphate-reducing bacteria.
2) A minimum negative shift in the measured IR-free potential of the structure of 0.100 V on application of the CP. That is the shift in the potential which is free from the potential drop resulting purely from the high resistivity of the environment.

16.4.2   Cathodic Protection Design

As far as it is practical, structures in contact with the ground should be electrically isolated from adjoining structures, which are not in direct contact with the ground. In the case of buried pipelines, for example, isolating joints or flange kits should be installed at the first flange above ground where the pipeline is connected to surface facilities, such as at processing sites, pumping stations, etc. [13]. Surface facilities usually incorporate a network of copper earthing systems to minimise the likelihood of electric shock and/or fire ignition risk. If not electrically isolated from the CP system, the copper earths will result in a high current drain from the CP system and may dramatically reduce the effectiveness of the CP of the structure to be protected. This problem can be very acute where a network of buried piping is being cathodically protected on a congested process site. In such circumstances, it is usually impractical to ensure that the buried piping is fully electrically isolated from the surface facilities. In addition to the problems caused by the copper earthing, the existence of shielding effects from concrete foundations can have an important influence upon the spread of protection on the buried piping. Thus, on such sites full protection of the buried piping in accordance with the standard protection criteria is rarely possible and a compromise is often the best that can be achieved. In such circumstances, a number of current drainage surveys using temporary CP equipment will often prove beneficial in determining the details of the final CP design and its likely overall effectiveness.

16.4.2.1   Groundbed

In order to optimise the amount of current available for CP of buried structures, it is necessary to provide an anode groundbed comprising an array of individual anodes connected in parallel to the positive terminal of the transformer rectifier. Historically, the most common anode material used for this purpose has been silicon iron, with 14.5% silicon, cast in the form of thin slender rods. Small additions of chromium (typically 4.5%) are normally included where the soil contains high levels of salt, such as in coastal regions. As with impressed current systems in environmental waters (see above), mixed metal oxide anodes have been used in place of silicon iron anodes in groundbeds in recent years.

Impressed current groundbeds can be constructed in a number of ways to minimise their electrical resistance and maximise current output as follows:

1) Where the structure is symmetrical, such as a storage tank, for example, as a series of individual vertical anodes surrounding the structure to be protected.
2) As a straight line of vertical anode rods with a uniform distance between each rod, all buried at a shallow depth.
3) As a straight line of horizontal rods, laid end to end also at a shallow depth.
4) As a straight line of vertical rods, one above the other in a deep well configuration.

The location and specific choice of groundbed design will be determined by the variation in soil resistivity with depth and the geometry of the facilities being protected. In all cases the anodes will be surrounded by a carbonaceous backfill, to increase the effective size of the anodes and their longevity, and reduce the anode groundbed to soil resistance.

For individual anodes described in (1) above, the resistance of the anode to backfill and the backfill to earth can each be determined from Eq. (16.1) in which ρ is the resistivity of the backfill and the surrounding earth, respectively. The difference between the two values of R is the electrical resistance between the anode rod and the outer edge of the backfill and this will determine the available current for a given applied voltage.

In the case of (2), the resistance of several anodes in parallel, R_p is given by [14] in Eq. (16.5) as follows:

$$R_p = \frac{0.00521\rho}{NL}\left(2.3\log_{10}\frac{8L}{d}-1+\frac{2L}{S}\left(2.3\log_{10}\left(0.656N\right)\right)\right) \qquad (16.5)$$

where:

ρ = the electrical resistivity of the backfill material (or earth)
L = the anode length in feet
d = the anode diameter in feet
N = the number of anodes in parallel
S = the anode spacing in feet.

For (3), the resistance of several anodes arranged in one horizontal line, the equivalent equation is as described by Eq. (16.6):

$$R_L = \frac{0.00521\rho}{L}\left(2.3\log_{10}\frac{4L^2+4L\{S^2+L^2\}^{\%0}}{dS}+\frac{S}{L}-\frac{\{S^2+L^2\}^{\%0}}{L}-1\right) \qquad (16.6)$$

where L and d are the same as for Eq. (16.5), and S is twice the depth of the horizontal anode bed from the ground surface in feet [13].

Deep well groundbeds comprise a series of anodes suspended one above the other down a deep vertical hole drilled for the specific purpose and supplemented with compacted low resistivity backfill. Overall depths can be as much as 500 ft (150 m), where soil conditions and the complexity of the structures being protected dictate. Deep well groundbeds are particularly beneficial when applying impressed current CP to buried structures on congested sites, such as at process plants, refineries, tank farms, etc. as a more uniform spread of protection is achievable here than with the alternative designs previously described. As the liberation of gas from the anodes occurs during operation, each deep well construction needs to incorporate a perforated vent pipe throughout the active part of the groundbed to minimise the risk of 'gas blocking' and premature loss of functionality. The electrical resistance of a deep well groundbed can be determined by considering the active section as single vertical anode and using Eq. (16.1).

In order to identify suitable locations for low resistance groundbeds and determine accurate soil resistivities to be used in groundbed design, extensive soil resistivity surveys are required at prospective ground bed locations [15].

16.5   Cathodic Protection of Well Casings

Water, oil, and natural gas well boreholes are usually stabilised using steel well casings. Each well will comprise a series of concentric steel casings, reducing in diameter and increasing in length with the depth of the borehole. Each casing is usually cemented in place, but often due to practical difficulties the cement is of limited effectiveness. Thus, depending upon the geological characteristics of the ground through which the borehole has been drilled, the externals of the casing may be at risk from corrosion and ultimately perforation due to intimate contact with corrosive ground conditions. Where from detailed soil corrosivity measurements or previous experience, the ground conditions are determined to be corrosive, it is customary to apply CP to the externals of the near surfcae well casings [12]. For the CP to be effective, there must be electrical continuity between the different casings at the wellhead. It is often considered beneficial to apply a coating to the casing externals to lower the current demand for effective protection. The coating must be mechanically robust to withstand the mechanical impact and shear forces exerted upon it during casing installation, to ensure that a significant portion of it remains.

While the application of CP to the externals of the well casing is analogous to that of a buried pipeline, monitoring of the effectiveness of the CP system on the well casing is far less straightforward. A detailed survey can only be carried out once the production tubing has been removed from the well. Using a 'corrosion protection evaluation tool' (CPET), comprising two contact electrodes a fixed distance apart, the magnitude of the CP current flowing up the inner casing can be determined. The recorded current-depth profile should reveal a uniform and progressive increase in current flow with decreasing depth if the casing is well protected. Any significant departures from this relationship will indicate that the casing is not well protected, external corrosion may well be occurring and adjustments to the CP system are required [16]. In addition, the use of E-Log i plots, where E is the potential measured against a copper/saturated copper sulphate reference electrode at the well head, and i is the applied current from the CP system can be used to determine the minimum average current density required to protect the casing from external corrosion without the requirement to remove the production tubing [15].

16.6 Cathodic Protection and AC Interference

Where buried pipelines run parallel to, or cross either overhead or buried AC power lines, then there is a need to consider the risk from AC-induced corrosion or lightning strikes. While accelerated corrosion of the pipeline may occur, there may also be safety risks for pipeline personnel working in the vicinity. In addition, damage to the pipeline coating, any insulating devices, and the cathodic protection equipment itself may result. AC interference may also prevent accurate monitoring of the effectiveness of the cathodic protection during routine cathodic protection surveys, pipeline maintenance, etc. Where AC interference situations arise or there are serious risks of lightning strikes, a full investigation of the extent of the interaction should be carried out by an experienced CP specialist. The investigation may necessitate the use of computer modelling techniques to identify the magnitude and geographical extent of the interference and to determine the corrective measures that are necessary to minimise any damage caused [17, 18].

16.7 Inspection and Testing

All CP systems on structures in contact with the ground should be installed with suffi-cient test facilities to allow the operator to check the operation and effectiveness of the CP system at regular intervals throughout the life of the structure. With pipelines, for example, above-ground test posts providing cable contact with the pipeline should be installed typically every kilometre to allow the routine measurement of the pipe to soil potentials at adequate time intervals. A correction to the measured potential for IR drop resulting from the resistance between the pipe and location of reference electrode - is usu-ally required [19]. Direct electrical contact with the pipe via the test posts also enables more detailed surveys, such as close interval potential (CIP) and direct current voltage gradient (DCVG) surveys to be carried out less frequently, but as and when required to reveal the full status of the corrosion protection being achieved. It may also prove advantageous to bury test coupons and permanent copper/saturated copper sulphate reference electrodes adjacent to the protected structure at selected locations to verify that sufficient current is being provided in the event that damage to the protective coat-ing has exposed the underlying steel.

16.8 Internal Cathodic Protection Systems

Steel process vessels, tanks and compartments containing an appreciable level of water during operation, such as oil-water separators, heater treaters [20], deoxygenation ves-sels, water storage tanks [21, 22], etc. may have CP installed to protect the wetted inter-nals from corrosion. On offshore platforms, the internals of flooded compartments may also be protected in the same way. CP of these water-wetted internals may or may not be in conjunction with a protective coating. Sacrificial anode CP systems are likely to prove more effective in low resistivity waters and where the internals of the equipment are coated, due to the low current demand in these situations. Conversely, impressed current systems are more appropriate where current demand is higher, that is where large uncoated areas need to be protected such as in water tanks, but are applicable where both high and low resistivity waters are being handled. If CP is to be applied in flooded compartments which are entered only infrequently, then some form of ventila-tion should be provided to prevent the build-up of hydrogen gas and avoid a risk of explosion.

The design of internal CP systems follows very much the same principles as men-tioned in Section 16.2 on CP in environmental waters, except that the designs need to take account of the limited opportunity for inspection and monitoring of the CP sys-tems outside of the operators' scheduled plant and inspection programme.

16.9 Summary

The topic of CP in relation to mitigating external protection of pipelines and struc-tures is briefly discussed. A brief reference to internal protection is also given. Effective implementation of CP measures is highlighted with a focus on practical aspects briefly outlining key parameters in design, choice of impressed vs sacrificial, implications of deepwater and issues in relation to alternating current (AC). The subject is extensive and here only a brief overview is given, underlining challenges for the operation and where appropriate references to further information is indicated.

16.10   Terminologies

The definitions of common terms are now given.[2]

16.10.1   Polarisation

Polarisation is the change of potential from a stabilised state, e.g. from the open-circuit electrode potential as the result of the passage of current. It also refers to the change in the potential of an electrode during electrolysis, so that the potential of an anode becomes more noble, and that of a cathode more active, than their respective reversible potentials. Often accomplished by the formation of a film on the electrode surface.

16.10.2   Calcareous Deposit

A calcareous coating is a layer that contains a mixture of calcium carbonate and magnesium hydroxide deposited on surfaces that are cathodically protected against corrosion, due to the protected surface's increased pH adjustment.

16.10.3   Open Circuit Potential

Open circuit potential (OCP) refers to the difference that exists in electrical potential. It normally occurs between two device terminals when detached from a circuit involving no external load.

16.10.4   IR Free Potential

Instant-off potential refers to a standard CP measurement method. It is the polarised half-cell potential of an electrode taken immediately after stopping the CP current. This potential closely approximates the potential without an IR drop (i.e. the polarised potential) when the current was on. Instant-off potential represents an effective on-potential (with IR-drop compensation). This potential can be used for the measurement of CP of a buried pipeline in the oil and gas industry.

Notes

1 Copper/saturated copper sulphate reference electrode ($Cu/CuSO_4$) in certain circumstances is used instead of Ag/AgCl. The former is commonly used for buried structures where environmental salinity is not too high whereas the latter is used in salt water

environments: high salinity can affect the stability of $Cu/CuSO_4$ reference electrode. The difference in measured potential between the two is small: $-0.85\,V$ (v $Cu/CuSO_4$ equals $-0.80\,V$ (v $Ag/AgCl$).

2 All adopted from https://www.corrosionpedia.com.

References

1 DnV, Cathodic protection design. DnV-RP-B401, 2010.

2 NACE, Corrosion control of submerged areas of permanently installed steel offshore structures associated with petroleum production. NACE-SP0176, 2007.

3 MTD, Design and operational guidance on cathodic protection of offshore structures, subsea installations and pipelines. MTD Limited Publication 90/102, 1990.

4 U.S. Military, Military specification: anodes, sacrificial zinc alloy. US Military Spec A-18001K, 1991.

5 NACE, Metallurgical and inspection requirements for cast sacrificial anodes for offshore applications. NACE-SP0387, 2014.

6 R A Adey (ed.) Modelling of cathodic protection systems. NACE 38442, 2005.

7 J Burke and C L Ribardo, BP America Thunder Horse-materials, welding and corrosion challenges and solutions. Offshore Technology Conference, Houston, Texas, 3–6 May, 2010.

8 ISO, Petroleum, petrochemical and natural gas industries-cathodic protection of pipeline systems, Part 1-On land pipelines. ISO 15589-1, 2015.

9 NACE, Corrosion control of underground storage tank systems by cathodic protection. NACE-SP0285, 2011.

10 CEN, Cathodic protection of buried metallic tanks and related piping. BS EN 1636, 2004.

11 NACE, External cathodic protection of on-grade metallic storage tank bottoms. NACE-SP0193, 2016.

12 NACE, Application of cathodic protection for external surfaces of steel well casings. NACE-SP0186, 2007.

13 NACE, Electrical isolation of cathodically protected pipelines. NACE-SP0286, 2007.

14 Peabody, A.W. (2001). *Peabody's Control of Pipeline Corrosion*, 2e (ed. R. Bianchetti). National Association of Corrosion Engineers.

15 ASTM, Standard guide for using the direct current resistivity method for subsurface investigation. ASTM D6431-99, 2010.

16 Morgan, J. (1987). *Cathodic Protection*. NACE International.

17 R A Gummow, A/C interference guideline final report. June, Canadian Energy Pipeline Association, 2014.

18 NACE, Measurement techniques related to criteria for cathodic protection on underground or submerged metallic piping systems. NACE TM0497-2012, 2012.

19 NACE, Mitigation of alternating current and lightning effects on metallic structures and corrosion control systems. NACE SP0177, 2014.

20 NACE, Internal cathodic protection (CP) systems in oil treating vessels. NACE SP0575-2007, 2007.

21 NACE, Galvanic anode cathodic protection of internal submerged surfaces of water storage tanks. NACE SP0196, 2015.
22 NACE, Impressed current cathodic protection of internal submerged surfaces of steel water storage tanks. NACE SP0388 2014.

Bibliography

Ashworth, V. and Booker, C.J.L. (1986). *Cathodic Protection: Theory and Practice*. Ellis Horwood.

Kermani, B. and Chevrot, T. (eds.) (2012). *Recommended Practice for Corrosion Management of Pipelines in Oil and Gas Production and Transportation*, Publication No. 64. The Institute of Materials, European Federation of Corrosion.

17

Corrosion Risk Analysis

The need to determine and ideally quantify the level of in-service risk is a key step allowing the safe functioning and mechanical integrity of all materials of construction and supporting corrosion control measures. This is an integral part of meeting compliance with health, safety and environment (HS&E) requirements and legislation. Therefore, risk assessment needs to be rigorously and systematically conducted, and periodically reviewed and revised, and to be present from the design stage, and maintained through commissioning, operation, and decommissioning.

A fundamental understanding of the range of corrosion mechanisms, their rates, and physical consequences, under favourable conditions, is a key starting point to safe and sound materials selection and in-service operational life. Respective corrosion threats are the subject of several chapters in this publication and no further description of these are included in this chapter.

The extensive use of carbon and low alloy steels (CLASs) due to their favourable cost, availability/supply, engineering properties and track record is always a strong draw despite the threat of corrosion being an ever present potential Achilles' heel. However, a shift to selective use of corrosion resistant alloys (CRAs) does not come entirely corrosion risk-free. Their use requires as much detailed consideration and justification as the use of CLASs and respective rigour when conducting a corrosion risk based assessment (RBA).

The use of RBA to identify credible corrosion threats and expressing them in terms of probability or likelihood versus consequence of failure (the Boston Square risk matrix)[1] is a pivotal element and process in developing a fit-for-purpose (FFP) corrosion management strategy and applied corrosion control programmes.

In addition, organisational structure and its operation are also key factors in the effective management of risks to plant integrity, and should not be lost sight of or underplayed. A useful way of viewing and reviewing the FFP of the organisational structure in place is through what is commonly termed the '4 P's' – People, Process, Plant and Performance. The importance of getting this side of the risk management equation right should not be underestimated or allowed to drift: it can easily undermine what is otherwise a technically sound corrosion risk assessment.

The elements of risk, its assessment, and treatment within the context of prioritisation as part of RBA are discussed in the present chapter. The topic is a primary purpose

Corrosion and Materials in Hydrocarbon Production: A Compendium of Operational and Engineering Aspects, First Edition. Bijan Kermani and Don Harrop.
© 2019 John Wiley & Sons Ltd. This Work is a co-publication between John Wiley & Sons Ltd and ASME Press.

behind the broader remit of Integrity Management (IM) dealt with in Chapter 18. IM is now firmly established as the all-embracing process for setting the requirements relating to risk, health and safety, and environmentally sound operations.

17.1 Risk

In general use, *risk* can be defined as [1]:

- *Noun* – hazard, danger, chance of loss or injury; the degree of probability of loss, a person, thing or factor likely to cause loss or danger.
- *Transitive verb* – to expose to risk, endanger; to incur the chance of an unfortunate consequence by some action.

Having identified a potential risk, it requires describing and quantifying in terms of probability/likelihood/frequency of occurring and consequence in order for it to be reduced to a safe and cost effectively managed level [2].

For risk involving corrosion, the principal focus lies in identifying the credible corrosion threats present and all the mitigating factors/barriers in place and their effectiveness expressed, most commonly, as a frequency or likelihood (probability) of a failure resulting. The consequence of a failure will largely be determined by the resulting failure mode – e.g. a pit leading to a weep versus a crack leading to a burst and rapid/catastrophic release of a pressure-contained environment – and design and operating circumstances.

The probability of failure is commonly expressed as the frequency of a failure event per year where frequency is a function of 1/life. Life as a function of corrosion can be simply expressed by the relationship between actual (measured) wall thickness and the wastage mechanism and mitigation effectiveness determining the prevailing mode of attack and rate. While the former is solely the domain of inspection, the roles of inspection and corrosion monitoring overlap to varying degrees in quantifying the latter. It should also be remembered that certainly inspection and often monitoring are lagging indicators as measures of corrosion damage present and rate. This is discussed further in Chapter 10.

17.2 The Bow Tie Concept

In line with process hazard analysis (PHA) [3], a commonly used approach for identifying potential hazards, threats, and consequences is the bow tie concept. One side of the bow tie identifies the hazard and the preventative controls (barriers) in place to mitigate an undesirable event; the other side of the bow tie identifies the consequences and required recovery controls – e.g. reassessment of barrier performance requirements; change of and/or additional barriers, frequency of monitoring and inspection; repair/replace actions; operational changes. This approach provides a basis for a systematic approach to identifying credible corrosion threats and the required barriers to effectively manage them through undertaking a detailed corrosion RBA for all defined elements and circuits that make up an operating system.

By way of example, consider a hypothetical multiphase CLAS trunk pipeline where, having conducted a corrosion RBA, the credible internal metal loss corrosion threats are: general corrosion, localised corrosion, preferential weld corrosion, microbiologically induced corrosion, flow induced corrosion, erosion-corrosion, and under deposit corrosion. The undesirable event ultimately is loss of primary containment. The *preventative controls* (barriers) side of the bowtie consist of: corrosion allowance built into the pipe wall, continuous corrosion inhibitor injection and regular batch treatment with biocide, minimum and maximum velocity limits, and regular running of a cleaning pig. The performance of barriers is monitored by: ensuring corrosion inhibitor injection rate is set to always deliver the required optimum concentration in the aqueous phase, regular analysis of water samples (pH, residual inhibitor concentration, dissolved iron concentration – Fe count – as ratio against manganese concentration to link the Fe source that of the steel pipe, planktonic bacteria), cleaning pig frequency plus pig trash analysis, functioning of well sand screens and presence of solids in separator, corrosion and erosion monitoring data, and inspection data especially from the running of intelligent pigs.

The *recovery controls* side, depending on the nature of the failure, may well include reassessment of the currently used corrosion inhibitor, leading to a change of dosage (injection rate) and/or change of product; review of the reliability of injection facilities and improvements to the injection management process. Any failure may well also call into question the continuing suitability – type, sensitivity, location, frequency – of the existing corrosion monitoring and inspection undertaken.

Describing, detailing, and quantifying the consequences side of the bowtie go beyond solely the remit of the corrosion engineer. The corrosion engineer will generally not have all of the information or necessary expertise required to make a detailed consequence assessment; this will require a multi-disciplinary team. For example, consequence assessments related to Safety and Environmental impact need to involve the Process Safety community within an organisation, forming part of Process Safety reviews – e.g. hazard and operability (HAZOP) studies used as part of a quantitative risk assessment (QRA). Likewise in considering the business consequence of a failure of equipment or system, a member of the related Operations Team (e.g. pipeline engineer, processes engineer, operations manager) needs to be part of the multi-disciplinary team.

17.3 Risk Matrix

In assessing the consequence of a failure, it will have elements of impact on HS&E and Business; and, depending on severity, can also impact the Licence to Operate (LTO) and reputation. How these elements are handled individually and collectively to give an overall consequence severity rating is usually governed by company policy and processes set down for managing risk. For example, the prescribed practice to be exercised by the risk assessment team may be that the highest consequence identified for each of the above elements for a given threat is ascribed for the purpose of plotting on a corrosion risk matrix.

Figure 17.1 shows a common form of risk matrix used typically to underpin the strategy, detail, conduct and update of a corrosion control programme for managing a particular threat – here solely for addressing corrosion under insulation (CUI). The numbers in Figure 17.1 defines the strategy to be applied, including minimum scope for managing CUI risk present with 1 the lighest and 3 the lowest risk. For example: Strategies 1 and 2 (highest risk locations) call for regular General Visual Inspection (GVI) at susceptible areas and Close Visual Inspection (CVI) at highly susceptible areas including sample removal of insulation and 100% removal of insulation typically every ca. 10 years' service; Strategy 3 (the lowest risk locations) is primarily reliant on regular GVI and CVI and the

Strategy		CUI probability				
		Very High	High	Medium	Low	Very low or unlikely
Consequence of failure	Very high	*1*	*1*	*1*	*1*	Review on case-by-case basis
	High	*1*	*1*	*1*	*2*	
	Medium	*1*	*1*	*2*	*2*	
	Low	*2*	*2*	*3*	*3*	
	Very low	*Repair on failure*				

Figure 17.1 A typical risk matrix for CUI.

use of Non-Intrusive Inspection (NII) techniques in determining a need to selectively remove insulation. In this example, probability considers several factors: substrate condition (cf. wall thickness), temperature exposure range, presence of temperature cycling or heat tracing, and exposure (i.e. inside module, outside, exposed to water deluge system, coastal site, heavy industrial site) which are scored for different substrate materials to give a probability ranking for each. In this example the numbers define the strategy to be applied, including minimum scope for managing the CUI risk present.

CUI has been chosen purposely to also emphasise the commonality and overlap of approach and application of corrosion RBA with that used for conducting risk-based inspection (RBI) [4–6]: Figure 17.1 has applicability to both. Arguably, the now seamless relationship that exists between the two has resulted in no small part from the pre-eminence of IM and its purpose, i.e. to continuously reduce operational risk by impacting the consequence and/or likelihood of premature failure of facilities, equipment, and infrastructure over the operating life of an asset.

17.4 Corrosion RBA Process

As stated earlier, risk assessments need to be rigorously and systematically conducted, and periodically reviewed and revised; and to be present from the design stage, and maintained through commissioning, operation, and decommissioning. It is therefore *essential* to have a robust RBA process that ensures comprehensive identification of all credible corrosion threats and consistent application and quantification of their severity. As importantly, this also includes the level of mitigation resulting from the presence and ongoing performance of all barriers and control measures in place.

The development of a robust RBA process is not a trivial exercise, as can be appreciated by merely drawing up a list of 'Standard Corrosion Threats', both internal and external, as the starting point. For upstream oil and gas operations the list readily exceeds 30 types of corrosion threat – covering the various forms of metal wastage and cracking – albeit only some will be relevant (credible) to a particular facility and/ or operating conditions. Each threat then needs to be expanded to detail which material is susceptible to it, under what conditions/applications, the anticipated damage morphology, expected rates of unmitigated attack, and suitable barriers and/or process controls and their expected effectiveness. The level of detail and values applied, including escalation and inhibition factors, will be drawn typically from: the open literature (e.g. standard reference books, technical journals); joint industry programmes; in-house funded testing/R&D; in-house and industry operating experience, lessons learnt and accepted industry 'best practice'; internal and national/international standards.

Fortunately, and again following from the pre-eminence of IM, it is not necessary here to have to start completely from 'ground zero' as far as the process per se is concerned. There is a growing number of commercially available service providers and software that in part at least offer process and framework as well as expertise, resources, and data management and reporting, for undertaking structured risk assessments. They can readily be found by conducting an internet search under terms such as Integrity Management, Asset Integrity Management, and Plant Integrity Management. However, whatever route is taken, it is important to understand the basic elements and their relationships that constitute an effective RBA process; and to ensure these are soundly incorporated and managed within any off-the-shelf commercial system adopted.

The corrosion RBA process consists of three principal activities – input, analysis, and output – which are discussed in the following sections.

17.5 Corrosion RBA: Input

Input consists of two primary steps:

- system data/information (intelligence) gathering;
- system segmentation.

The former is concerned with gathering *all* relevant data and information to satisfactorily enable a sound and full completion of the corrosion RBA. The latter is concerned with dividing a system into discrete segments – corrosion circuits – each of which is potentially exposed to common threats and mitigation methods. This step is key to managing the complexity of strategy, performance and reporting required to consistently deliver successful corrosion control throughout a system.

Exercising due diligence with no complacency in the fullness of undertaking these two steps is paramount to ensure working with a comprehensive base from which to develop a sound RBA and consequently a fit-for-purpose corrosion management strategy and ongoing control programme. Working through these two steps will likely take considerable time and effort and certainly require keen attention to detail and accuracy. After the first pass RBA, it should become slicker and quicker with each review, giving attention to step changes in, for example, process conditions, system modifications, safe operating limits resulting from progressive corrosion (e.g. reduction in the maximum allowable operating pressure [MAOP] for a pipeline). Often harder to spot or appreciate early, but just as important and inevitably critical is the presence of creeping changes.

Key sources of system data/information are:

- basis of design (BoD); piping and instrumentation diagrams (P&IDs); process flow diagrams (PFDs): Isometric drawings;
- operating conditions and history;
- materials specs – metallic and non-metallic materials, including coatings, seals, gaskets, etc.;
- MAOP and other safety-related operational considerations (e.g. does a corrosion circuit contain safety critical equipment and should these be considered separately rather than part of a circuit?);
- corrosion monitoring and inspection data;
- cathodic protection system design and performance surveys;
- chemical treatment/deployment history;
- failure history (of system being assessed and similar elsewhere).

The quality of available system intelligence will also hinge heavily on having the right people in the room when working through the two steps of the RBA input. It is not the sole task of the corrosion engineer to undertake and complete. The corrosion engineer will need to engage with other relevant disciplines – e.g. process engineering, pipeline engineering – who collectively agree on system data/quality and relevance and segmentation. It is also important to have people in the room who physically know the system being assessed and how it's *actually* being operated which will likely involve operators and technicians. And as a final comment, care should be exercised in placing reliance *solely* on use of P&IDs, PFDs, etc. as fully representing the *current* build and operation of a system.

17.6   Corrosion RBA: Analysis

Three steps form the core of this activity:

- identifying the credible corrosion threats present;
- identifying the barriers and control processes in place and their effectiveness;
- determining degradation rates and failure modes.

These steps are applied systematically to each corrosion circuit/component identified and detailed in the precursor Input activity.

Identifying the credible corrosion threats present will draw on the list of standard corrosion threats referred earlier: the list should be drawn up and documented ahead for reference and consistency in undertaking this step but will be subject to periodic review. The list should cover internal and external exposure conditions considering corrosion, erosion, and different types of environmental cracking.

Having rigorously established the *credible* corrosion threats present for each corrosion circuit/component, the resulting severity and mode of attack/failure presented by each credible threat and, where possible, rates of attack then need to be assessed and detailed and quantified as far as possible. Then the consequent level of mitigation resulting from the effectiveness of barriers and control measures in place needs to be applied to the latter. Again, here there is a need for consistency of approach and in defining and detailing approved sources of data, information, guidelines, models, etc. to be drawn on: the aforementioned standard corrosion threats reference document should include these against each threat. The level of success here will depend on integrating *all* the elements likely to conspire in making a corrosion threat *credible*; but just as importantly how that corrosion threat will in fact show itself.

17.6.1 An Example: Flowline Corrosion

Figure 17.2 shows what is commonly termed bottom-of-line groove corrosion; here in an onshore CLAS wet crude oil flowline. This form of attack has also been experienced in a number of offshore flowlines, and is sometimes called tramline corrosion where two distinct parallel lines of attack are present, at least in the early stages. The primary source of this form of metal loss corrosion is the presence of an aqueous phase acidified by dissolved CO_2 and so prediction of the rate using one of the available CO_2 corrosion models [7] would be an appropriate first step as discussed in Chapter 5. However, care needs to be exercised as clearly there is more to it than that.

The pipeline had always been operated under stratified flow, resulting in a continuous aqueous phase running along the bottom-of-line between the 4 and 8 o'clock position with some occasional swirling due to production upsets. Furthermore, there was a notable level of solids – a mix of sand particles and corrosion product – present predominantly in the aqueous phase. Modelling showed the aqueous phase to be potentially scaling in terms of precipitation of corrosion product ($FeCO_3$) resulting in surface filming under the low flow conditions (typically $<1.5\,\mathrm{m\,s^{-1}}$) able to afford some degree of protection. However, this would be continuously disrupted by a rolling movement of solids being transported along the bottom-of-line in the separated

Figure 17.2 An example of bottom of line groove corrosion. (see colour plate section).

aqueous phase: this may also adversely affect the corrosion inhibitor performance too. Nevertheless, the presence of solids was not enough to constitute an erosion (mechanical) threat to the bare steel substrate but was able to potentially catch out the actual relevance of a simply modelled predicted corrosion rate!

Figure 17.2 is a good example of how a combination of physical and (electro) chemical factors actually determines the resulting morphology and rate of corrosion damage experienced. Clearly a lack of appreciation (or awareness) of the full picture would likely lead to a far less severe risk assessment of the situation until it was too late! The actual corrosion rate associated was in fact higher than that predicted simply using a CO_2 corrosion model. This raises another consideration – viz. the need to apply an escalation or adjustment factor taking account of the form of attack, here being highly localised. Typically this can result in applying a multiplier of two to three to the predicted rate based solely on fluid corrosivity – without taking due account of any added influence resulting from a particular corrosion morphology and any time-dependent and often highly specific mechanistic and physical factors. The multiplying factors are typically set by drawing on past experience – in-house and industry-reported – addressing similar situations, relevant research work, and targeted lab testing, and correlating in-service measured with predicted (modelled) corrosion rates. However, they may inevitably have an element of subjectivity that results in such factors varying in magnitude between operators. The corrosion engineer may well have to make a judgement call having reviewed all the relevant facts and evidence versus the general and specific capabilities of corrosion modelling: this is part of reaching a balanced assessment of the risk. This also highlights the importance of ensuring the active and passive corrosion barriers in place function as expected.

The need to apply an escalation or adjustment factor is often required where localised corrosion is a threat, pitting in the presence of H_2S [8] perhaps being that most commonly encountered, which can occur at levels below that normally classed as sour (≥ 0.0034 bara P_{H2S}) as discussed in Chapter 6. A similar requirement exists in other situations such as in the presence of preferential weld corrosion (PWC) and mixed metal galvanic corrosion. The need for an even higher multiplier may be required where top-of-line corrosion (TLC) is deemed to be a credible threat in wet gas lines, especially if H_2S and volatile organic acids (primarily acetic acid) is present. Furthermore, the severity of such conditions is extremely sensitive to corrosion inhibitor selection and subsequent deployment/system management – i.e. concentration and consistency of injection – and to internal cleanliness of a pipeline.

In the absence of any direct modelling capability of a credible corrosion threat, assessing the risk is then more often about ensuring conditions do not prevail for the threat to initiate at all. This may mean at the design stage having to select a higher (more-resistant) alloy depending on the application. Threats here include pitting and crevice corrosion, especially although not exclusively associated with CRAs, and cracking mechanisms such as stress corrosion cracking (SCC) again more commonly associated with CRAs, but not exclusively.

17.6.2   An Example: Sulphide Stress Cracking

A potential threat that can affect all but the most highly alloyed materials commonly used in the oil and gas industry is that of sulphide stress cracking (SCC) if there is credible likelihood of H_2S being present in produced fluids and gas at any time during the life of field. As discussed in Chapter 6, there is the requirement for full compliance with ISO 15156 [9]. However, operator internal standards and practices may call for additional further refined conformance measures, such as application of internally generated domain diagrams – pH_2S v pH – for specific alloys [10] and often tied to specific or preferred suppliers. It should be noted that the susceptibility to SSC is potentially highest at ambient laboratory temperatures so it is common for domain diagrams to be generated under such 'worst case' conditions. However, what complicates the picture is the unexpected presence of H_2S – often having been assessed as highly unlikely at the BoD stage – occurring at some point during the operating life of field and therein the (growing) risk it then represents. Designing for sour service per se incurs an additional cost on a project's CAPEX (capital expenditure), which invariably brings added pressure in considering the likelihood of SSC ever becoming a credible threat. Commercial considerations can bring additional complexity to setting acceptable practical levels of H_2S for continuing safe use of a facility or pipeline not originally designed for sour service and especially where contracted to handle third party oil and gas: a sound knowledge of service history and current condition has heightened criticality.

17.6.3 Localised Attack

A further complication in assessing the risk associated with localised forms of attack, such as pitting and crevice corrosion and the various forms of cracking, is an associated induction or incubation period for the necessary localised conditions to establish and conjointly interact – this period could be hours, days, months or years. This places further importance on upfront materials selection and maintaining service conditions below defined threshold values – e.g. a material is not allowed to experience conditions outside its defined safe operating limits or domain. Drawing solely on test and field data available in published technical and product performance literature can be conservative as a general guide and therefore requires undertaking selective lab qualification testing. However, trying to meaningfully accelerate induction times under laboratory conditions may in itself introduce risk and uncertainty in applying to actual field conditions! Nevertheless, the judicial use of general 'go-by' parameters such as pitting resistance equivalence number (PREN), critical crevice temperature (CCT) and critical pitting temperature (CPT) has a useful part to play, whereas sound application of a coating and/or cathodic protection can often mitigate the risk of localised attack and may even remove it completely.

17.7 Corrosion RBA: Output

Three steps form the core of this activity:

1) Determining a mitigated likelihood (probability) of failure (typically resulting in loss of primary containment) occurring within each corrosion circuit resulting from each of the credible corrosion threats present.
2) Assigning a consequence of failure (severity level) to the latter to enable plotting on a common risk matrix to give a clear statement of the individual and relative significance each threat present.

3) Producing a set of actions to inform and update/refine the corrosion management strategy and control programmes in continuously reducing corrosion related risks to an acceptable level.

The first two steps involve a high element of judgement even where a 'quantitative assessment' of the mitigated rate of attack has been derived from modelling, inspection, and monitoring data and preferably a combination thereof. Models clearly have their limitations and may well rely on the use of broadly derived empirical factors to better account for the rate of certain localised forms of attack – e.g. PWC, ToLC, presence of H_2S, flow-assisted corrosion, whereas monitoring and inspection data are generally highly location-specific – typically presenting a < 1% view of a system and certainly in the case of weight loss coupon and inspection data functioning as lagging indicators. This highlights the critical importance of ensuring a strong interface between corrosion RBA and RBI programmes and the intelligence they generate on the ongoing condition of operating systems.

Where a mitigated corrosion rate can be determined or reasonably estimated, remaining life can be calculated based on the actual thickness measured during the most recent inspection, less the minimum wall thickness required by the appropriate design code allowable stress due to pressure and mechanical and structural loading. From the resultant remaining life, the probability of failure – typically expressed as the likelihood of failure within the next year – can be calculated. However, the need to conduct a more rigorous fitness-for-service assessment (as detailed in API 579 [11]) and/or Engineering Critical Assessment (ECA) [12] may be required as wall loss becomes more extensive, or deeper or complex in morphology.

The judgement element becomes even more acute where dealing with credible threats that by their nature can only be assessed in terms of susceptibility – i.e. a requirement for conditions not to exceed a single or set of threshold conditions often specific to a material (e.g. environmental cracking such as SSC and SCC). Here close scrutiny of operating conditions and their management history to-date versus the BoD will be a critical consideration. Reference to the performance of similar facilities and systems with similar operating conditions within an organisation and that reported by other operators can also provide a useful reality check. However, for such non-time-dependent threats, the probability of failure comes down to engineering judgement. Here having in place a probability descriptor guide to draw on is essential, typically covering an incremental range of the lowest of $<10^{-6}$ per year (e.g. an event unlikely to happen within two to three times design life) to the highest of ≥ 1 per year.

Since the performance and hence actual effectiveness of barriers often have a strong operator element to them, this too needs to be factored in when allocating a likelihood of failure. An example here is the operation of water injection systems and how well the level of residual dissolved oxygen in the injection water is being continuously maintained below the typical threshold limit of 10–15 ppb.

The allocation of a likelihood and consequence morphing into a credible corrosion threat should involve joint input by and agreement between the corrosion engineer and system-relevant operation disciplines – e.g. process engineer, pipeline engineer. In addition, in the case of consequence, it should be led by appropriate representation from an operator's process safety/HS&E community. Consequence is a complex outcome to have to assign a single severity level to. There is a need to consider impact of a failure with respect to HS&E, and Business (financial and non-financial). How these are factored together will likely be a matter governed by adherence to strict company policy and guidelines. However, for day-to-day corrosion management purposes, the corrosion risk matrix used is that based primarily on HS&E consequences. It is then common to assign the highest consequence identified for a specific threat.

17.8 Corrosion RBA: Overall Process

Figure 17.3 summarises the whole corrosion RBA process, resulting in a set of actions to inform and update or refine the existing corrosion management strategy and programmes for each corrosion circuit; and set the detail of the corrosion control and monitoring programmes in operation – the plan, do, check and adjust working cycle. It provides the probability/likelihood of the occurrence of the credible corrosion threats identified to enable a collective risk matrix to be produced for each corrosion circuit or facility. The matrix succinctly shows the absolute and relative severity and criticality of each threat and provides a ready means of seeing how effective a corrosion management programme is functioning – i.e. how each threat moves over time absolutely and relatively on the matrix. It might be useful to view the health of the risk matrix as the beating heart of the overall corrosion management process and its effective function. Furthermore, as already discussed, there is a strong interface with the setting and conduct of RBI programmes. The darkest shaded squares in Figure 17.3 (normally coloured red), designate the highest risk/highest priority domain (of Boston Square of the risk matrix); next highest shaded grey squares (normally amber) designate the high risk domain;

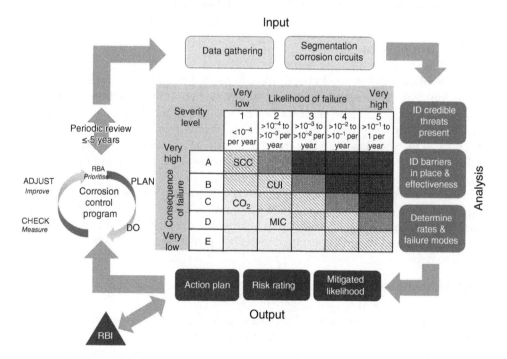

Figure 17.3 A summary of the whole corrosion RBA process.

lightest shaded squares (normally yellow) designate medium risk domain; and unshaded squares designate low risk domains. The Boston Square axes will be determined primarily by regulatory and company HS&E requirements and performance conformance, with the consequence axis rating including a business impact factor (cf. as discussed in 17.1).

It may then be helpful to prioritise work plans based on addressing the highest risk items, e.g. those that have the most impact on risk reduction, where there are large differences between the mitigated and unmitigated corrosion rate, particularly in the absence of any design or nominal corrosion allowance. It may also help to view the identified corrosion risks versus manageability – a simple high, medium and low measure of ability to consistently and continuously mitigate/eliminate the risk.

17.8.1 Ensuring Continual Fitness for Service

Finally, it should be remembered that conducting the corrosion RBA process is part of assuring a system's continuing fitness-for-service, i.e. in its current condition, is it capable of safely operating at defined operating conditions for a defined operating period (typically a one year rolling review window). The need to refresh the corrosion circuits' RBAs will be affected by: the measured/monitored actual performance and conduct of the corrosion control programmes; significant changes in operating conditions; occurrence of major upsets and near misses; evidence that remnant life is reducing faster than anticipated; extension of required operating life versus original design or past anticipated remaining life. Nevertheless it is good practice to refresh all corrosion RBAs every five years irrespective; and it may be prudent to move to a shorter interval for ageing assets. Turnarounds (TARs) and planned and unplanned shutdowns may equally warrant or present an ideal opportunity for refreshing the corrosion RBAs.

17.9 Risky Business

Having in place a well-defined RBA process supported by appropriate reference documentation, with the latter subject to regular review to ensure its continuing currency, are key requirements. However, where dealing with risk, it should be recognised that in itself it can be a 'risky business' to conduct due to uncertainty and inadvertent subjectivity creeping in, when landing a risk ranking. It is therefore informative and valuable to reflect on what others in a broader view and appreciation of risk have to say that offer insights for the corrosion engineer to be cognisant of.

In his book, *Risk Intelligence*, Evans [13] refers to a special kind of intelligence for dealing with risk and uncertainty – cited as that common to weather forecasters, professional gamblers, and hedge-fund managers. If to be believed, it is something that is necessarily not a natural or well-taught expectation of the competencies of a corrosion engineer.

17.10 Behaviours

For the corrosion engineer, a useful parallel with where the RBA process sits in his/her daily activities can be made with reference to what Kahneman [14] refers to as System 1 and System 2 approaches to judgement and choice. System 1 operates automatically and quickly as a response, and more typically reflects that of a corrosion engineer dealing with the immediate day-to-day business of front-line incidents, operational upsets, request for guidance, etc. System 2 concerns attention to demanding mental activities such as complex computations and the subjective experience of agency, choice, and concentration. It is a System 2-type response where conduct of the corrosion RBA process clearly best resides. The operations of System 1 can generate surprisingly complex patterns of ideas, but only the slower System 2 can construct thoughts in an orderly series of steps. However, this is not to say System 2 doesn't have a front line role to play too.

There is a 'what you see is all there is' rule [14] which also has some resonance in conducting risk assessments, especially where working with limited evidence and data and information (intelligence). In particular, biases of judgement can often unknowingly creep in such as:

- *Over-confidence* – through a failure to allow for the possibility that evidence that should be critical to a judgement is missing.
- *Framing effects* – resulting from different ways of presenting the same information evoking different reactions: an individual normally only sees one formulation.
- *Base-rate neglect* – losing sight of the significance of statistical facts in base data and information, even if limited in amount, especially when first considering a situation.

Where assigning a probability (likelihood) is a key element of a process, that itself can introduce an element of risk resulting from the level of uncertainty associated with the assigned probability. Furthermore, this conditional probability will be affected to a greater or lesser extent as new/more evidence is acquired or a significant event – for example, pH_2S or temperature exceeds a specified safe threshold limit for a short period of time – occurs.

17.11 Bayes' Theorem

Re-evaluating probabilities as additional data/information (intelligence) is gathered is a common consideration in probability theory and where Bayes' Theorem [15] – concerned with conditional probability – comes to the fore. It shows that simply adopting a refined probability based only on new evidence and completely ignoring (discounting) the prior probability can be misleading as exemplified in Table 17.1.

Bayes' Theorem and Bayesian analysis have to date not featured strongly, if at all, in the routine conduct of corrosion RBA but it is clear this will likely need to change in the

Table 17.1 Simple example showing application of Bayes' Theorem.

Prior probability – initial estimate of given corrosion threat likely (%)	**5**			
Fresh/new information/intelligence acquired				
Probability conditional on being directly relevant/related/certainty of being true (%)	**50**	**70**	**90**	**99**
Probability conditional on not being directly relevant/related/ certainty of being true (%)	**50**	**30**	**10**	**1**
POSTERIOR PROBABILITY – Revised estimate of given corrosion threat likely (%)	**5**	**11**	**32**	**94**

future in undertaking conditional RBAs. Useful introductory discussion of Bayes' Theorem and its application in context of risk intelligence and living with uncertainty and the art and science of prediction are presented in references [13, 16], respectively.

Bayesian reliability concepts are being used to estimate the evolution of corrosion defects in pipelines [17]. Ainouche [18] describes how Bayesian modelling of the kinetics of corrosion in a gas pipeline combined with the data resulting from only one in-line inspection enables a credible evaluation of the risk of failure. Sabarchim and Tesfamariam [19] demonstrate the use of a Bayesian belief network-based, probabilistic, internal corrosion hazard assessment approach for oil and gas pipelines: multiple corrosion models and failure pressure models have been incorporated into a single flexible network to estimate corrosion defects and associated probability of failure.

17.12 Moving Forward

Looking ahead, arguably RBA is an area ripe for the use of neural networks and Artificial Intelligence (AI) to play an increasingly major role. It seems likely that Bayesian networks will have a key part to play [20], ably supported no doubt by algorithms to better manage speed of reaction to events and changing circumstances and conditions, and to ensure consistency and efficacy of application of the corrosion RBA process.

17.13 Summary

The potential risks to the ongoing integrity of plant, facilities, and operating systems presented by the various forms of corrosion can be identified, quantified and effectively managed down to safe residual levels through rigorous and systematic application of an RBA process, built around a consistently applied set of sequential steps. While the corrosion RBA process arguably represents an intuitive straightforward and common-sense approach, success is founded on having the required detail and methodology underpinning each step clearly set down as a precursor and reference source. The latter is not a trivial undertaking but once in place will significantly smooth subsequent consistent applications of the RBA process. It will be necessary to refine the details and methodology from time to time as lessons are learnt and better understanding of corrosion processes, safe material operating limits, new and improved corrosion models, etc. become established.

Finally, it is worth also reflecting on *knowing what you know* and *what you don't know can hurt you* [13, 15]: you've religiously worked through the corrosion RBA process so *must* have all the bases covered, right! And then there are infamous *unknown unknowns* [13] or perhaps better viewed as the *corollary of mistaking the unfamiliar for the unlikely* [15].

Note

1 A Boston Square risk assessment matrix is a project management tool that allows a single page – quick view/priority of the probable risks evaluated in terms of the likelihood or probability of the risk and the severity of the consequences.

References

1 Chambers (2006). *The Chambers Dictionary*, 10e. Chambers Harrap Publishers Ltd,.
2 ISO, Risk management: principles and guidelines. ISO 31000, 2009.
3 Sutton, I. (2015). *Process Risk and Reliability Management*, 2e. Elsevier.
4 API, Risk based inspection. API RP 580, 2016.
5 API, Risk based inspection technology. API RP 581, 2016.
6 DnV, Risk based inspection of offshore topsides static mechanical equipment. DNV-RP-G101, 2010.
7 R Nyborg, CO_2 corrosion models for oil and gas production systems. NACE Annual Corrosion Conference, Paper No. 10371, 2010.
8 SN Smith and MW Joosten, Corrosion of carbon steel by H_2S in CO_2 containing oilfield environments – 10 year update. NACE Annual Corrosion Conference, Paper No. 5484, 2015.
9 NACE, Petroleum and natural gas industries – Materials for use in H_2S-containing environments in oil and gas production. ANSI/NACE MR0175/ISO 15156, 2015.
10 MB Kermani, PJ Cooling, JW Martin, and P Nice, The application limits of alloyed 13%cr tubular steels for downhole duties. NACE International Conference, Paper No. 98094, 1994.
11 ASME, Fitness For Service (FFS) Assessment Standard. API 579–1/ASME FFS-1, 2007.
12 R Kuprewicz, A review, analysis and comments on engineering critical assessments as proposed in PHMSA's proposed rule on safety of gas transmission and gathering pipelines, Pipeline Safety Trust, 2016.
13 Evans, D. (2012). *Risk Intelligence: How to Live with Uncertainty*. Atlantic Books.
14 Kahneman, D. (2012). *Thinking, Fast and Slow*. Penguin.
15 Bayes, T. (1763). An essay towards solving a problem in the doctrine of chances by Thomas Bayes (published after Bayes death by friend Richard Price with multiple amendments and additions). *Philosophical Transactions of the Royal Society of London* 53 (0): 370–418.
16 Silver, N. (2013). *The Signal and the Noise*. Penguin.
17 Bisaggio, H.C. and Netto, T.A. (2015). Predictive analyses of the integrity of corroded pipelines based on concepts of structural reliability and Bayesian inference. *Marine Structures* 41: 180–199.
18 A Ainouche, Future integrity management strategy of a gas pipeline using Bayesian risk analysis. 23rd World GAS Conference, Amsterdam, 2006.
19 Sabarchim, O. and Tesfamariam, L. (2016). Internal corrosion hazard assessment of oil and gas pipelines using Bayesian belief network model. *Journal of Loss Prevention in the Process Industry* 40: 479–495.
20 (2015). A Ananthaswanny, New Scientist – Chance: The Science and Secrets of Luck, Randomness and Probability. In: *Putting Chance to Work*, 213–232. Profile Books Ltd.

18

Corrosion and Integrity Management

Integrity Management (IM) has become firmly established as the principal cross-discipline methodology for developing the integrated requirements to address mitigating all asset integrity-related losses. This is achieved through a structured and performance-managed approach, including operational processes and best practices, standards (national, international and company), expectations and responsibilities, and the setting of performance metrics. IM is primarily targeted at ensuring delivery of health and safety and environmental (HS&E) sound operations is consistently achieved; and mitigating any knock-on consequences to business performance, reputation and ultimately the Licence to Operate (LTO).

This bigger picture is a natural overarching home for Corrosion Management (CM), as historically many integrity-related losses and near-miss incidents have been corrosion-related, either directly or as a result of poorly integrated operational practices.

The IM/CM relationship has undoubtedly strengthened the importance and visibility of and attention given to CM at the working level; arguably far more so, at all levels within an organisation, than merely referencing the cost of corrosion, as outlined in Chapter 1.

Chapter 17 highlights the need to determine and ideally quantify the level of in-service risk to the safe functioning and mechanical integrity presented by the corrosion performance of materials of construction. This sits at the heart of the current chapter to defining a fit-for-purpose CM strategy and detailing sound corrosion control programmes and performance measures.

18.1   Integrity Management (IM)

IM is a subject for a book in itself and detailed coverage is beyond the scope of the present chapter. It is important to appreciate the wider role of IM in managing all hazards, from whatever source, that have the potential to adversely affect the safe and sound operation of all operating elements of an asset; and in turn may indirectly or directly affect the potential for corrosion to occur.

IM is not a standalone discipline. It is an integrated set of applied activities, as a continuous assessment process, to assure all equipment and facilities are designed,

Corrosion and Materials in Hydrocarbon Production: A Compendium of Operational and Engineering Aspects, First Edition. Bijan Kermani and Don Harrop.

procured, constructed, installed, operated, maintained, and eventually decommissioned to avoid accidents due to loss of primary containment, fire, or structural collapse. Importantly, IM is about continuously reducing operational risk by addressing the consequence or likelihood of premature failure of facilities, equipment and infrastructure over the life-cycle of an asset.

When considering the safe and optimum utilisation, up-time and life-cycle costs of facilities and equipment, a key factor is maintaining operational availability – commonly addressed by reliability, availability and maintainability (RAM) modelling, studies and management systems. However, safe and sound ongoing system integrity will affect the 'availability on demand' of facilities and equipment in which corrosion has the potential to impact all.

Most countries have legislation that requires an operator to have a policy, commonly enshrined within a company standard or equivalent, on the prevention of major accidents and environmental damage associated with hydrocarbon production, processing, and handling operations. The legislation may well contain regulations that are prescriptive and dictate specific activities and schedules which must be complied with as part of an IM programme. This can result in some initial conflict that will require resolution before bringing a system, facility, etc. into operation and may even be an important consideration at the design stage.

18.1.1 Overview of IM Elements and Practice

There is no international standard that defines what constitutes the minimum elements and content of an IM process and plan. However, there are guidelines targeted at specific systems/facilities, for example, DNV-RP-F116 Integrity Management of Submarine Pipeline Systems [1], or Energy Institute Guidelines on Integrity Management of Subsea Facilities [2]. Furthermore, conducting an internet search on IM will result in identifying a number of service providers offering their developed process and software, expertise, and experience.

In practice, it is common in the oil and gas industry to have an overarching IM document setting out the minimum requirements and aims to be met, including operator/company values and expectations relating to risk, HS&E performance, reputation and LTO. This leads to the development of specific processes and details for day-to-day application and IM performance management. This will include the setting of key performance indicators (KPIs) for defined operating areas, for example: Wells Integrity Management System (WIMS), Pipelines Integrity Management System (PIMS), or Facilities Integrity Management System (FIMS).

The overarching document will typically set out definition and scope, minimum requirements, and performance management, addressing the following:

- hazard evaluation and risk management;
- facilities and process integrity;
- protective systems and emergency response;
- management of change.

Underpinned with detail and requirements covering:

- competence and accountabilities;
- practices and procedures;

- incident investigation and learning;
- performance management.

Of particular importance is identification of all safety critical equipment (SCE) – assuring their availability and continuing sound functionality will secure the required risk reduction of major incidents and hazards.

18.1.2 Risk and Hazard Evaluation

Adopting a systematic methodology for identifying *all* credible hazards present in order to assess the risk they present (probability and consequence, as outlined in Chapter 17) is key to the success of an IM programme. This drives the IM process summarised in Figure 18.1. The core goal is to *continuously* reduce operational risk which can become increasingly challenging to manage as an asset ages and especially in the event of a changing business imperative for service life extension beyond original design life.

One of the key outputs will be Boston Square risk matrices, (more detailed discussion and an example are given Chapter 17) that can be used to readily view the relative risk-ranking of a range of hazards and setting priorities and manageability.

There are a number of hazard evaluation and risk assessment techniques in common use, with some suited to particular applications. For example, in order of increasing detail: hazard identification (HAZID); hazard and operability studies (HAZOP); layer of protection analysis (LOPA); failure modes and effects analysis (FEMA); quantitative risk assessment (QRA).

18.1.3 Implementation

Once the detail of an IM programme has been developed, it is important to measure its implementation and success through performance metrics made up of a suitable balance between leading and lagging indicators. KPIs should be identified and agreed

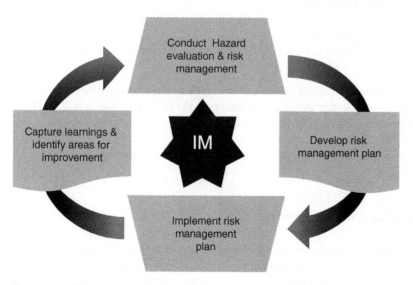

Figure 18.1 The IM process.

by all stakeholders and used to track and regularly assess and report IM performance. Performance metrics are a key feed into Quarterly and Annual Performance Reviews (QPRs/APRs) and reporting, and the setting of annual engineering plans (AEPs). They provide a ready window for timely identification of areas for improvement. Leading metrics could include the number of integrity-related actions due versus those completed; corrosion rates which will lead to a loss of containment. Lagging indicators could include the number of high potential incidents (HIPOs); number of incidents resulting in loss of hydrocarbon containment.

IM and CM are primary frontline processes in assuring equipment and facilities remain fit-for-service (FFS): in their current condition are capable of operating safely at defined operating conditions for a defined operating period. FFS is commonly subject to annual review but will depend on the severity of the operating conditions, historical performance, and incidents record, and age.

Successful implementation also depends on all parties connected with delivery of an IM and CM programme knowing their respective roles and responsibilities and how they inter-relate. A tool used in project management that has seen adoption here is a RACI matrix or chart. The acronym RACI stands for:

- *Responsible:* as a rule, this is one person with responsibility for the development and execution of the IM and/or CM programme: a principal role of the corrosion engineer. This may include delegation of specific tasks and actions as appropriate.
- *Accountable:* the one person accountable for the correct and thorough completion of programme plans and actions: who responsible is accountable to. The accountable person will depend on operational size and complexity and management structure but, for example, could be the Engineering Authority, IM Manager or Operations Manager. This also could be the corrosion engineer where they delegate specific tasks.
- *Consulted:* the people who provide specialists information, input and guidance with whom there is two-way communication, for example, subject matter experts and engineering technical authorities.
- *Informed:* the people who need to be kept informed of progress, particularly where KPIs are affected by the outcomes. Who this includes will depend on the size and complexity of an asset and the management structure, so, for example, it could be the Asset Manager.

Creating a RACI matrix involves listing all the tasks and activities on one axis and all the relevant roles on the other axis to successfully deliver the IM/CM programme. Then complete the cells accordingly, ensuring every task has a role responsible and a role accountable for it. No tasks should have more than one role accountable. The RACI matrix and any subsequent modifications and changes must be agreed by all the shareholders before use.

18.2 Corrosion Management (CM)

The process for the development and application of a pro-active CM programme mirrors that of the basic blueprint for the IM process shown in Figure 18.1. Its objectives may be simply stated as 'to prevent incidents and unplanned losses due to identified credible corrosion threats; and to optimise life cycle approach to equipment integrity affected by corrosion, including FFS.'

The starting point is undertaking a rigorous and systematic corrosion risk assessment. It is here that the facilities and equipment and physical boundaries to be covered by the resulting CM programme are defined. This will include operating envelopes, materials of construction, defining corrosion circuits, and listing all corrosion barriers (active and passive) in place to manage the credible corrosion threats identified. The corrosion risk based assessment (RBA) process and its role as a principal element of a holistic CM programme are considered in more detail in Chapter 17.

Figure 18.2 summarises the primary elements and activities that complete the CM programme cycle, building on a plan-do-check-adjust framework that has traditionally driven the working process but now giving equal importance to RBA. While the need to undertake risk assessments has always been recognised, in the past it has tended to be submerged along with other considerations in the detail, making it all too easy to skip through in a desire to 'get into action' driven by reaction to events.

The CM programme needs to be a living process that is strategically shaped and tactically driven by proactive management of *all* the credible corrosion threats identified and mitigated through the operational controls and the active and passive corrosion barriers in place. It is important that the risks the credible corrosion threats continue to present – viewed both in terms of likelihood and consequence (risk matrix) – are regularly reviewed and revised as necessary, as learning and new/more evidence result, and therein to continuously improve the CM programme. Steady and step changes in operating conditions, but also significant upsets and transient changes, including

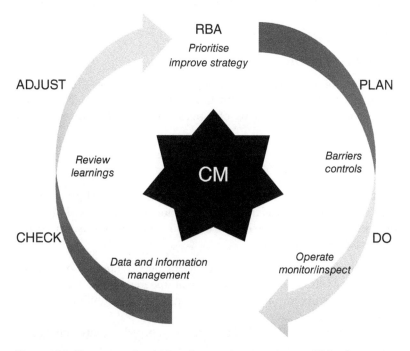

Figure 18.2 Elements and activities of a corrosion management (CM) programme.

during start-ups and shutdowns, should prompt assessing the need to update individual corrosion threat assessments. As mentioned in Chapter 17, creeping changes can be particularly difficult to pick up or appreciate their significance early enough but can potentially end as being very damaging. This is where undertaking an external review of the CM programme can prove invaluable to the on-going success of the programme to be typically scheduled a minimum of every five years. Nevertheless, it is good practice to refresh *all* corrosion threat assessments every five years. This should take into account strategic changes in the future service requirements of the operating unit/facility (e.g. any major shift in required operating life; anticipated changes in field operation and performance) versus current FFS and possible adoption of new technology.

The CM programme may also involve an element of prioritisation considering factors such as: manageability versus performance metrics, running versus replacement cost, remnant life versus life extension, and current operational resources to run the programme (highlighting a need to recruit and/or reallocate resources).

The *do* and *check* steps, that feed the *adjust* step, are where the daily pulse of the CM programme resides. In today's information technology (IT) world it is possible to generate a great deal of data – both direct (e.g. corrosion monitoring and inspection) and circumstantial (e.g. flow rate, pH, presence of solids) – much in real time that singularly and collectively indicate and/or measure how well a CM programme is performing. Performance and data management are consequently key activities in the smooth running and overall effectiveness of a CM programme and are discussed further in Sections 18.2.1 and 18.3.

Defining roles and responsibilities is also of key importance to the success of a CM programme and where the RACI tool discussed earlier is an effective way of addressing this.

18.2.1 Performance Management

An important component of a CM programme is performance management, where all the data are brought together and reviewed to ensure that the objectives of the programme are being met. Typical elements of a performance management system are described below.

18.2.2 Performance Indicators (PIs)

Performance Indicators (PIs) are measures of all the elements of the CM programme. There are often a large number of PIs as they need to include, and are based on, individual pieces of equipment and processes. PIs are used on a day-to-day basis by corrosion engineers and include items such as:

- *actual versus planned activity:* e.g. number of coupons pulled, number of maintenance pig runs, number of inspections, risk assessment updates;
- *barrier monitoring:* e.g. cathodic potential values, inhibitor availabilities, coating degradation values;
- *corrosion rates:* e.g. coupon values, electrical resistance probe values, ultrasonic (UT) probe values, calculated vs measured rates;

- *process variables:* e.g. temperature, pressure, flow rates, fluid composition, periods of shutdown, etc. These parameters are extremely valuable for capturing small, incremental changes that, over time, can add to a significant change to the original design basis – known as 'creeping change'.
- *inspection results*: total cumulative wall loss, wall loss since last inspection.

PIs are often built into a traffic light coloured format so that items which need addressing are coloured red and items meeting targets are coloured green: orange is also used sometimes to draw attention to a possible deteriorating performance measure. An example is given in Table 18.1 for a sea water injection system that also includes sulphate removal to mitigate the risk of barium sulphate scale formation in the reservoir and especially around injection well and production well perforations.

If all items are always coloured green, the value of the scorecard reflecting PIs should be challenged to determine if more appropriate indicators are needed. For example, should certain programmes be stopped or the criteria made more stringent?

18.2.3 Key Performance Indicators (KPIs)

KPIs are a small number of indicators that summarise performance of the overall corrosion management programme – typically three or four which may be a distillation of the much larger number of PIs. These are targeted at company managers and their primary purpose is to highlight areas of concern so that help can be provided on a timely basis for intervention. A common mistake is to try and show managers all of the PIs and to focus on where things are going well rather than on problems and where help is needed.

18.2.4 Performance Reviews

It is important to regularly review the PIs to determine if the objectives of the programme are being met. The frequency needs to be sufficiently high so that any issues can be addressed in a timely manner. Issues include developing negative trends in performance against PIs, occurrence of significant operational events (e.g. system upsets/downtimes; corrosion probe malfunction) and HIPOs that may require intervention or tactical revision to the CM programme. For a typical oil and gas facility, the frequency of ongoing performance meetings would usually be monthly. There may also be a need to feed into the IM quarterly performance reviews (QPRs) depending on the nature of any corrosion-related incidents in the period and the IM PIs/KPIs.

The primary purpose of Annual Performance Reviews (APRs) is to review the performance of the previous year and to set the targets for the next year. This may also involve reviewing the overall objectives of the CM programme with company management to ensure the basis for them still applies. Feed into the IM APRs may also be required, and if not directly affecting the AEPs, to have awareness of their content as it might impact the CM programme.

It is easy for corrosion engineers to be consumed with the day-to-day running of their corrosion programmes and not find time to step back and look at the programme as a whole. Bringing colleagues from another operation or using external consultants to peer review the programme is an excellent way to get an independent check on the

Table 18.1 IM dashboard: sea water injection system (see colour plate section).

Monitoring/Mitigation		Week#12	Week#13	Week#14	Frequency/Target	Corrective action
Hypochlorite injection, Continuous, SW Lift Pump	Status PMvT				Continuous @ 20 ppm	
SRB/GAB (Planktonic) u/s Multimedia Filter	Status/PMvT		N/A		Bi-weekly/10 SRB per sample & 100 GAB ml^{-1}	
Chlorine, u/s Multimedia Filter	Status/PMvT		N/A		Daily/0.5–1 ppm	
Fe, u/s Multimedia Filter	Status PMvT	0.2 ppm	N/A	0.2 ppm	Bi-weekly/Flat Trend	
TSS, u/s Multimedia Filter	Status PMvT	10 mg l^{-1}	No sample	5 mg l^{-1}	Daily/ <3 mg l^{-1}	
Cl$_2$/Bisulphite (OS)Continuous in VDAT sump	Status PMvT	3 ppm	1 ppm	5 ppm	4 ppm when SRU operating	
DO, Orbisphere d/s of VDTA	Status PMvT		Offline	12 ppm	Online/≤10 ppb	
Orbisphere calibration	Status		Offline	???	Every 30 days	
Biocide treatment, u/s SRU	Status/PMvT	200 ppm	200 ppm	400 ppm	1 hour/3 days @ 200–400 ppm	
Bisulphite (OS) residual, u/s SRU cartridge filters	Status/PMvT				Twice daily/0.64–2.56 mg l^{-1}	
Cl$_2$, u/s of SRP Cartridge Filters	Status/PMvT				Twice daily/zero	
DO, CHEMetrics (manual) SRU product header	Status/PMvT				Twice daily/≤10 ppb	
SRB/GAB (Planktonic) SRU product header	Status/PMvT		N/A	No sample	Bi-weekly/10 SRB/sample & 100 GAB ml^{-1}	
SRB/GAB (Sessile) SRU product header	Status/PMvT		N/A	No sample	Bi-weekly/10 SRB/sample & 103 GAB ml^{-1}	
ER corrosion probe, d/s water injection pump				0.2 mm year^{-1}	Online/Running average ≤0.1 mm year^{-1}	
Corrosion coupon, d/s water injection pump & Last coupon	Status/PMvT		Coupon Change due Week # 50 0.07 mm year^{-1}, no pitting		Yearly/ ≤0.1 mm year^{-1}, nom pitting	
DO, CHEMets (manual) d/s water injection pump	Status/PMvT				Twice daily/ ≤10 ppb	
Water velocity - in any given pipe section in system	PMvT				2 < Velocity < 10 m s^{-1}	

Green	Orange	Red	N/A

TSS = total suspended solids; PMvT = performance-measured versus target; DO = dissolved oxygen. CHEMetrics is a company making a colorimetric DO test kit ampules charged with reactant that when filled with a sample of treated sea water colours according to the level of DO present. It is a quick visual means of measuring DO by comparing the colour of the tested water sample against a colour chart. (https://www.chemetrics.com/index.php?route=common/home). CHEMets are most commonly used in the upstream, being quick and simple to use albeit subject to a degree of user judgement on the colour change but a good means of correlation with the on-stream DO sensors. There may be other makers of such kits but the author is not aware of them.

OS = oxygen scavenger; SRU = sulphate removal unit; VDAT = vacuum de-aeration tower;

value and performance of the programmes. External review and assurance might usefully be timed to fit with the recommendation that the RBA behind each CM programme is fully reviewed at least every five years.

18.3 Data Management

One challenge that the modern corrosion engineer faces is how to integrate the abundance of data that are available. In the last decade tools have become available that provide a graphical interface for the integration of data. This facility is much more powerful than having a desk full of books, reports, and presentations containing the data. In this section a few examples are outlined but are by no means exhaustive and only serve as a precursor to other potential applications.

18.3.1 Outdoor Facilities

For outdoor facilities the Geographical Information System (GIS) technology is now widely used to display a digital map of facilities, such as wells, pipelines, storage tanks, etc. The map can then be overlaid with an infinite number of data sets that provide different disciplines with important information. For the corrosion engineer, such layers include corrosion monitoring and inspection data, an example of which is portrayed in Figure 18.3.

Although GIS will never replace the benefit of having a corrosion engineer who knows the geographical location, it does come close, as the GIS view is interactive, allowing the engineer to zoom in and out of specific locations. Displayed data points can be clicked-on to bring up more detailed information, such as the history of all inspections at the location and what techniques were used. There are no data which cannot be built into a GIS map and hence it is a very powerful tool.

18.3.2 Indoor/Enclosed Facilities

For indoor facilities the increased use of laser scanning to create a digital image of the equipment has been an exciting development. Modern facilities are usually designed using 3D drawings which can be used to create visual models of the facility. Unfortunately this does not apply to older facilities and, in some cases the 3D drawings never make it out of the Project and into Operations. In such cases 3D scanning can be used to create a model as shown in Figure 18.4. After the model has been created, it can be overlaid with data in a similar way to the GIS as shown in Figure 18.5.

18.3.3 Data Collation and Representation

As discussed in Chapter 10 on corrosion trending, it is not merely collecting data that is important but how it is then used and 'smartly' analysed to provide timely intelligence on the current condition of a system and flag up potential issues to enable a pro-active rather than a reactive response. Furthermore, it is not restricted to data resulting from what might be deemed direct corrosion measurements, but also that which might be

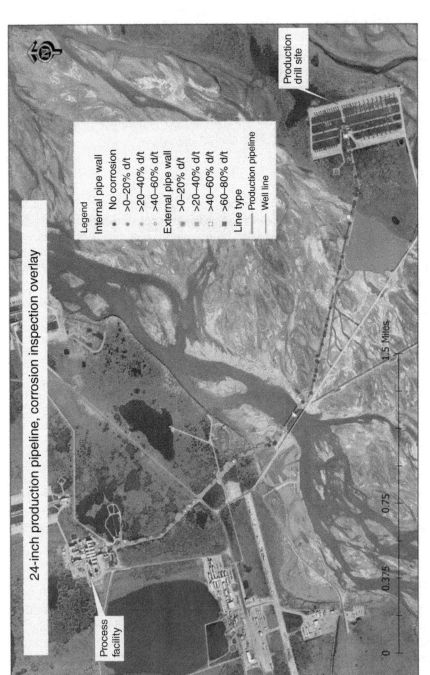

Figure 18.3 A GIS map of a pipeline showing inspection data. (see colour plate section).

Figure 18.4 A 3D model of a processing facility created using 3D laser scanning. (see colour plate section).

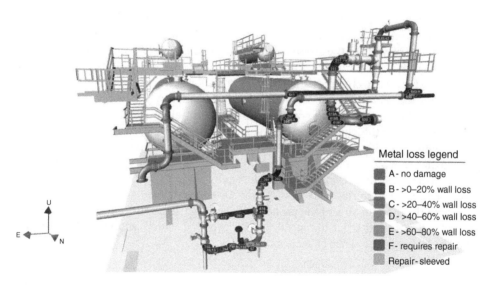

Metal loss legend

A - no damage
B - >0–20% wall loss
C - >20–40% wall loss
D - >40–60% wall loss
E - >60–80% wall loss
F - requires repair
Repair - sleeved

Figure 18.5 A 3D model of indoor facility piping showing inspection data. (see colour plate section).

termed as indirect or circumstantial. The former covers corrosion trending (monitoring and inspection data). The latter includes data such as: flow rate, oil/water/gas ratio, temperature and pressure, sand production, pH and water composition, dissolved gases (e.g. CO_2, H_2S, O_2), organic acids, microbiological activity, iron (Fe/Mn) counts, inhibitor residuals concentration, line cleanliness/cleaning pig frequency, meeting target chemical injection rates (e.g. corrosion inhibitor, oxygen scavenger, biocide), production chemical injection excursions, coating condition, CP potential surveys, to name but a few.

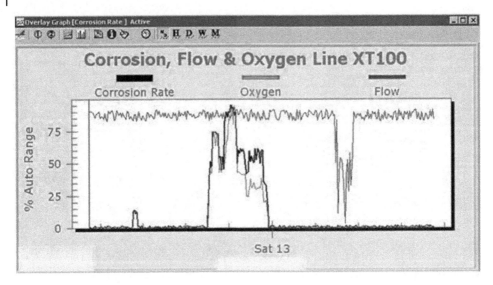

Figure 18.6 ER probe data from water injection line. (see colour plate section).

Many companies provide software that can take multiple data inputs and are able simply to correlate them with the corrosion monitoring data to produce a set of dashboard displays (see the example given in Figure 10.4). As an immediate visual appreciation of a situation, it is a valuable capability, an example of which is shown in Figure 18.6 for a sea water injection line with an LPR probe.

As outlined in Chapter 7, a basic understanding of the oxygen corrosion mechanism reveals that the rate is affected by dissolved O_2 concentration and flow rate. Depending on its magnitude, the change should result in a corresponding response by the LPR probe (with a possible slight delay). Corrosion modelling of the recorded situation may give further short- and longer-term insight into its significance especially factoring in the frequency of such an event. However, just as importantly, if the corrosion probe had shown no response to the spike in dissolved O_2 concentration, this might indicate an issue with the functionality of the probe. As flagged in Chapter 10, probes and coupons are not infallible and a flat-lining ER or LPR probe, if merely accepted at face value, could be giving a false sense of security.

18.4   The Future

The availability of large amounts of data should enhance decision making and the soundness of an IM/CM programme but the sheer amount of data, combined with the pressure to make smarter and faster decisions can overwhelm decision-makers. This holds true in the world of corrosion as much as any other field. However, the growing capability and expertise in using 'big data' are increasingly opening up the potential to get even smarter in data handling and analysis. For example, the ability to establish process signatures that are known to precede actual events in an asset's historical operating experience and build predictive models that alert to the presence of such signatures in future data that may present a threat to integrity.

This is where predictive data analytics has the potential to make a step change in the pre-emptive way CM (and IM) are conducted and their success – the ability to undertake real-time and historical scan statistics – pattern search, definition, and matching – with expert system type/rules modelling function. This potential has been recognised for some time [3] as a core element of the long-talked-about 'field of the future' or 'intelligent/smart facilities'.

Such developments also provide grounds for optimism in not getting caught out by 'unknown unknowns' as discussed in Chapter 17 on risk analysis. The use of heuristic analysis common to many computer anti-virus programs offers promise here and is being assessed by some operators. The vision discussed in 2008 [3] has arguably moved past being a 'nice to have' to become essential in achieving a step change in IM/CM.

There has been some notable progress through data-to-desk (D2D) and satellite telemetry technology making real-time, centralised, collaborative centres able to remotely access and support globally located operating sites 24/7. The centres serve as another set of eyes and source of advice. Data are transmitted in real time via a satellite link, and the field operators can see and talk and hold review meetings, etc. with their remote centre counterparts via an always-on video link.

18.5 Summary

IM has become firmly established as the cross-discipline holistic methodology for developing the integrated requirements to address mitigating all asset integrity-related losses. This is achieved through a structured and performance-managed approach including operational processes and best practices, standards (national, international and company), expectations and responsibilities, and the setting of performance metrics.

CM is a primary component of the bigger picture embraced by IM, where historically many integrity-related losses and near-miss incidents have been corrosion-related, and continues to be so.

At the heart of IM and CM are systematic risk and hazard assessments – typically expressed on a Boston Square risk matrix (probability/likelihood versus consequence) – used as a semi-quantitative tool to readily view the relative risk-ranking of a range of hazards and setting priorities versus manageability. This enables strategy and programme detail to be developed; the latter built around a plan-do-check-adjust cycle. The continuing effectiveness and success of the programme are then driven by having in place a robust performance and data management process – including the setting of performance and key performance indicators (PIs/KPIs) – and a review/reporting regime where all credible risks and hazards are fully re-assessed and refreshed periodically and at least every five years. The core goal is to continuously reduce operational risk which can become increasingly challenging to manage as an asset ages.

The availability of large amounts of data should enhance decision making and the soundness of an IM/CM programme. The growing capability and expertise in utlising 'big data' are increasingly opening up the potential to get even smarter in data handling and analysis to enable a stronger predictive and pre-emptive capability.

References

1 Det Norske Veritas AS, Integrity Management of Submarine Pipeline Systems, Recommended practice DNV-RP-F116, February 2015.
2 Energy Institute, Guidelines for the management of integrity of subsea facilities, April 2009.
3 D Harrop and W Durnie, Using modelling and data analytics to signal change: can we avert a 'midlife crisis'?, SPE 4th International Conference on Oilfield Corrosion, Aberdeen, 27 May, 2008.

Bibliography

Kermani, B. and Chevrot, T. (eds.) (2012). *Recommended Practice for Corrosion Management of Pipelines in Oil and Gas Production and Transportation*, EFC Publication Number 64. Maney Publishing.
Morshed, A. (2016). *An Introduction to Asset Corrosion Management in the Oil and Gas Industry*, 2e (e-book). NACE Publication (Product Number: 37619-E).

19

Corrosion and Materials Challenges in Hydrocarbon Production

The search for new sources of fossil fuel continues to move to deeper, hotter, harsher, and more corrosive operating environments. Furthermore, as energy demand grows globally, future challenges become more widespread and evident. While significant progress has been made over the years in understanding the root causes of corrosion threats and metallurgical aspects with advances in innovative solutions to address key challenges facing the sector, there still remain grey areas where in very isolated situations fit for service solutions fall short of expectations. In these rare occasions, design and deployment of metallurgical solutions and corrosion mitigation measures have proved inadequate, with undesirable consequences. The cause may have been implementation of inadequate integrity management and control systems.

This chapter sets out to outline future direction, reflecting energy sector outlook, describing existing challenges facing the hydrocarbon production industry sector, and briefly highlighting notable avenues where shortfalls have been experienced. The chapter views challenges imposed on the hydrocarbon production industry sector through two themes: (i) what the corrosion and material challenges are and underlines respective solutions which can make a significant impact; and (ii) where current knowledge of materials and corrosion has in rare isolated circumstances failed to address the appropriate solutions. On the former, high pressure/high temperature (HPHT) fields are outlined which continue to play a major role in the search for new sources of energy.

19.1 Energy Viewpoint and the Role of Technology

The general industry sector projection indicates that, for the foreseeable future, fossil fuel and in particular hydrocarbon will remain a principal source of global energy. The world economy is expected to almost double over the next 20 years, driven by emerging economies, with growth averaging 3.4% per year and, more than two billion people lifted from low incomes [1–4]. Meanwhile, the world's population is projected to increase by around 1.5 billion people to reach nearly 8.8 billion. Global demand for energy is set to continue to grow over the next two decades as prosperity increases and the world's population rises [1, 2]. However, the mix of fuels used will change, driven by technological advances and environmental concerns, and demand will grow more slowly than in the past as energy is used more efficiently. Carbon emissions will likely

Corrosion and Materials in Hydrocarbon Production: A Compendium of Operational and Engineering Aspects, First Edition. Bijan Kermani and Don Harrop.
© 2019 John Wiley & Sons Ltd. This Work is a co-publication between John Wiley & Sons Ltd and ASME Press.

continue to rise, but with an expectation of being slower than in the past. For the future, there is a real need for energy that is affordable, sustainable, and secure.

Growth in demand versus production of oil and gas is proving to be a difficult and increasingly complex and volatile picture to predict, certainly in the short to medium term of two to five years. Oil in recent years has become particularly affected by factors such as the growing use of alternative/renewable energy sources, the green agenda, regional conflicts and politics, and advances in technology recovery methods versus economics (e.g. fracking). The demand for gas, however, has generally become less adversely exposed and affected by the latter factors to become a more acceptable energy source, and substantial growth in gas production to meet an anticipated rising demand seems a sounder prediction. Delivering such demands to fuel human and economic developments necessitates the right technology with timely delivery to meet and transform business performance. Technology continues to play a fundamental role in the hydrocarbon industry sector's business success in which materials/corrosion technologies and related innovative measures are paramount. The sector still has an open attitude to a phased implementation of step-outs, game changers, and innovative avenues.

19.2 Future Focus Areas and Horizon

The growing world economy will require more energy, but consumption is expected to grow less quickly than in the past – at 1.3% per year over the period (2015–2035) compared with 2.2% per year in 1995–2015 [1] from 75 million barrels per day (mb/d) in 2000 to around $104\,\mathrm{mb\,day}^{-1}$ in 2040 with substantial growth in gas. As mentioned, technology remains fundamental to the sector's business success within which environmental and social responsibilities are interwoven with operational and financial responsibilities [1–3].

Increasing demands for hydrocarbons have led the industry to broaden its search on three distinct horizons:

- more recovery – from what was already found;
- more discovery – of conventional resources;
- more diversity – broadening the frame of where hydrocarbons are found, moving to hotter, deeper and more challenging areas.

All these face increasing challenges and require unique technologies for their development. These horizons generally indicate increasingly harsher conditions, some may encounter more H_2S due to a combination of factors which are outside the scope of this chapter.

19.3 Challenges in Materials and Corrosion Technology

It is apparent that technology continues to play a fundamental role for the oil and gas sector's business success in which development of materials and corrosion mitigation technologies and innovative measures are paramount. The importance of materials and corrosion technology in achieving safety and security, minimising the impact on the environment and reducing cost should always be borne in mind. In addressing such measures, there is a global move to an integrated subject of integrity management (IM), as discussed in Chapter 18.

Challenges in materials and corrosion technology are in relation to four principal themes including:

- improved metallurgy;
- cost reduction;
- innovation and design;
- improved corrosion resistance.

The respective materials and corrosion avenues primarily consist of:

- integrity management;
- rejuvenation and life extension;
- remote monitoring and inspection;
- HPHT and ultra HPHT (moving to harsher areas –further description below);
- harsher conditions (deeper, more remote, hotter, more corrosive, solid production);
- carbon capture, transportation and storage (CCTS);
- more appropriate and high performing chemicals;
- shift to gas production;
- increased production;
- virtual (unmanned) and low intervention facilities;
- untreated water injection;
- sand management/erosion.

Again, all with the focus placed on improved safety, minimising the impact on the environment and reducing cost.

19.3.1 HPHT Reservoir Trends

In the past, the challenging environments of HPHT wells were considered uneconomic, but as technologies and experience evolve, tapping these reservoirs has become a reality on an increasing basis. Typically, HPHT wells were not considered economically viable until the mid- to late-1990s. In fact, the term HPHT was only first coined within the industry in the mid-1980s [5–10].

HPHT developments are defined as developments of reservoirs with a pressure exceeding 69 MPa (10 000 psi) and a temperature above 150 °C (300 °F). This is the definition used by the Department of Trade and Industry [6, 9]. These classification boundaries are arbitrarily divided into several sub-categories mainly influenced by the stability limits of elastomeric seals, electronic devices, and common well service tool components. This arbitrary division is schematically shown in Figure 19.1 with boundaries for operating conditions in terms of current, short-/medium (2–5 years), and long-term (+5 years) also indicating domains of geothermal production scenarios.

19.4 Shortfalls in Technology Implementation and Knowledge Partnership

The economic impact of corrosion threats on upstream operations is discussed in Chapter 1, indicating that some 10–30% of this cost can be reduced by implementing currently available corrosion control practices [11]. While implementation of innovative technologies, methods, and measures together with reflection of lessons learnt are vital

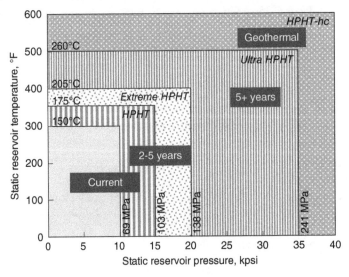

Figure 19.1 Arbitrary limits of HPHT well conditions.
The hybrid circuitry (hc) classification defines the most extreme environments – wells with temperatures and pressures greater than 260 °C (500 °F) or 241 MPa (35 000 psi). Such pressure conditions are unlikely to be seen in the foreseeable future in E&P. However, bottomhole temperatures (BHTs) in geothermal and thermal-recovery wells already exceed 260 °C.

in moving the industry forward, some limited recent unexpected and undesirable experiences have clearly demonstrated the respective challenges that this may impose, highlighting great care and scrutiny will be needed to implement innovations. In particular, it is vital that the optimum properties of components to meet demanding conditions are achieved by a combination of stringent manufacturing practices in steelmaking, mechanical and heat treatment processes, and finishing. Careful attention to these practices leads to the development of metallic materials with superior properties allowing a fit for service solution [12–25].

While limited, the industry experience of using a number of components can be summarised here as examples of where technology implementation can be erroneous with undesirable outcomes, some potentially costly. In some instances, it should be noted that the shortfalls are not purely due to inappropriate use of innovative technologies, but that the material was not up to the standards the designer anticipated or the materials/circumstances/environment that caused the failure had not been identified at the design stage. Furthermore, it is not to say that the failures were due to alloy/environment systems and the culprit may have been inadequate integrity management and control systems. Undoubtedly, there is no substitute for:

- taking care to fully understand the material/environment/circumstances combinations that are being considered, and these become increasingly complex and more varied when dealing with innovative technologies;
- ensuring that a tight materials specification is prepared in advance with all the necessary QA/QC to support it.

Once again, it should be emphasised that these adverse experiences are few and far between and not an adverse demonstration of or comment on the vital role the corrosion and material community has played and continues to play in providing safety, security, and protecting public welfare. Nevertheless, they are noted as lessons to be learnt in adopting innovations with undue care which may have significant consequences. The particular point is that in the majority of cases, due care in metallurgical processes and QA/QC were not implemented effectively. In addition, for materials selection, a need for rigorous evaluation of steady state as well as transient prevailing environmental conditions is a vital element of a methodical strategy.

19.4.1   25%Cr Super Duplex Stainless Steels

A few incidents of unexpected failure of super duplex stainless (SDSS) components have been reported in the past. These include corrosion cracking of production tubular used in the completion of HPHT wells and cracking of SDSS manifold piping [12–14]. In the case of the former, it was revealed that chloride stress corrosion cracking (Cl⁻SCC) had emanated from an area of hardened material with modified microstructure on the outside surface of the production tubular. Areas of high hardness and modified microstructure showing anomalous features had resulted from a combination of cold drawing fabrication anomalies and subsequent grinding. The source of high chloride was attributed to the initial annulus fill and the leak of calcium chloride from the B annulus, following which this chloride had concentrated within the cracks due to evaporation cycles.

For the manifold [25], the cause was due to hydrogen-assisted cracking attributed to a combination of unacceptably high local stresses (applied plus residual), inappropriate balanced metallurgical phases, and cathodic over-protection.

19.4.2   22%Cr Duplex Stainless Steel

Environmental assisted cracking of 22%Cr duplex stainless steel (DSS) pipework has been associated with non-anticipated change of environments during operation. Cracks were reported to have initiated from both inside and outside areas associated with corrosion by high salinity, low pH brine formed by extreme evaporation of produced water caused by a large pressure reduction step at high temperatures [15–16]. Cracks from the outside were attributed to chloride SCC failure mode of 22%Cr. The mechanism that initiated cracks from the inside was attributed to the evaporation of produced water which had created a brine with extremely high calcium and magnesium chloride concentrations. The resulting exotic brines had not been anticipated and were outside the range of environments for which the process materials were qualified.

Another case relates to the longitudinal cracking of high strength 22%Cr DSS tubulars which occurred on removal from the well [17]. This was attributed to galvanically induced hydrogen stress cracking (GHSC) potentially due to a combination of high strength and hydrogen uptake at elevated temperature exposure. It was generated through galvanic coupling with carbon steel casing, both exposed to the annulus fluid with specific chemistry that may have exacerbated hydrogen uptake. On removal from the wells, trapped hydrogen and residual stress may have led to the longitudinal parting.

Pitting corrosion of 22%Cr DSS manifold piping has also been attributed to inappropriate use of water during a prolonged hydrotesting period [14].

19.4.3 Alloy 718 and 725

A failure investigation on the root cause of a 718 tubing hanger failure concluded it was caused by the combination of a susceptible microstructure characterised by heavy delta phase precipitation at grain boundaries and the presence of hydrogen. The copper plating, used as anti-galling treatment, was mentioned as a possible source of hydrogen [18].

In a somewhat similar vein, hydrogen embrittlement of an alloy 718 forged casing hanger components in HPHT wells was associated with the formation of hydrogen due to the decomposition of the caesium formate brine. This intergranular failure occurred through a combination of an unfavourable microstructure, stress concentrations, and absorbed hydrogen [18–19].

These limited anomalies have led to the development of the API 6A 718 standard designed to prevent the use of materials having a microstructure susceptible to hydrogen embrittlement. This API standard describes the acceptance criteria for fabrication and thermo-mechanical treatments, mechanical properties and gives acceptable and unacceptable microstructures using typical photomicrographs.

Failure of a UNS N07725 (alloy 725) [20] completion component was attributed to hydrogen-assisted cracking caused by nascent hydrogen which emanated from a chemical reaction and highly localised stress caused the cracking to occur. The chemical reaction was associated with an improper thread cleaning and doping process. A number of recommendations [19] were given on the use of appropriate thread compounds to exclude MoS_2, cleaning procedure with care and other remedial measures to alleviate the threat in future operations.

19.4.4 Alloy 17-4PH

There has been circumstantial evidence of failures with UNS S17400 (17-4PH) through the years but the published accounts have been few. It is generally understood that the reason for such failures may have been due to misinterpretation of alloy limitations. Surpassing these limitations during operations may have led to the failures. These limits are documented in the literature albeit as differing and inconsistent values [21–23].

Reported failures of UNS S17400 have shown that in most cases the content of residual ferrite and austenite in this nominally martensite grade is a key issue governing its mechanical properties. These have led to some unusually high hardness areas beyond the anticipated values. Failures are normally associated with a combination of poor microstructure, inadequate heat treatment or high local stresses and to some extent low tolerance of this alloy of the anticipated environmental conditions [21–23].

19.4.5 Super-Martensitic 13%Cr Line Pipe Steels

Some limited experience of hydrogen-assisted failure of super-martensitic line pipe 13%Cr steel has been attributed again to a combination of environmental variables. These include deployment beyond the limits of application, over-protection by cathodic protection (CP), use of inappropriate welding consumables allowing high local hardness zones and/or lack of appropriate post-weld heat treatment (PWHT), and metallurgical parameters [24]. In addition, cracking from super duplex fillet weld on supermaternisitic 13%Cr pipeline was associated with a combination of local stress, strain, and hydrogen generated from CP.

19.4.6 Riser Systems

Risers are critical elements of all offshore production facilities. It is the final link in bringing live, non-stabilised, and invariably wet crude oil and gas at elevated reservoir temperature and pressure, on to a facility; and the subsequent export of stabilised/semi-stabilised crude oil and gas either into a transmission pipeline system or remote loading buoy facility for onward sale and processing. Some references to riser systems and their materials of construction are given in Chapters 14 and 15 and here, due to its challenging role, further description is considered necessary.

Located directly beneath or alongside the topsides floating or fixed leg processing facilities, any unexpected deterioration in mechanical integrity, be it detected or suspected, has the immediate potential to introduce an elevated safety risk to topsides operations personnel. Risers are almost always fabricated from CLASs, which may be internally clad with a CRA where produced fluids have very high inherent uninhibited corrosivity [24].

A riser is continuously exposed to the internal and external environments, both presenting ready conditions for corrosion to occur. Therefore, due attention to mitigating the various forms of threat that can occur need to be considered at the design stage and effectively managed throughout its service life. The threat of external corrosion occurring in the splash zone has received and continues to receive particular attention. Very few serious incidents, and certainly any associated loss of life, are known to have occurred. Most of these corrosion threats, if not all, have been external in origin. The low incident rates perhaps owe much to advances in inspection and external coatings technology together with the effectiveness of current corrosion inhibitors, if still not always helped by ease of direct access through design – e.g. use of riser bundles.

However, as a complication to the picture, there have been several cases in recent years where internal pitting in risers has been detected. Poor chemical treatment and/or management of hydrotest fluids being left in situ for longer than planned, or invariably a period of wet parking before hook-up/final installation have been attributed as the root cause. This has affected carbon steel risers and at least one alloy 825 internally clad riser, in all cases exposed to inhibited produced fluids, and all deployed in deepwater. Detection was through the first scheduled run of an in-line inspection (ILI) tool some five years after start. Having detected the pitting, it had proved difficult to quickly and unequivocally determine the root cause. Furthermore, it raised uncertainty about any contributory role played by the incumbent corrosion inhibitor used to treat the produced fluids. Most importantly, it proved hard to determine whether the pitting had been controlled by inhibition, especially not knowing from a single ILI run how long the pitting had been there and whether it was still active. Needless to say, such a situation results in a significant amount of highly focused activity offshore, in the laboratory, and in the office to satisfactorily and safely bottom out, arguably something resulting as a consequence of 'original sin'. On the positive side of the learning curve, in all cases it was shown that inhibitor treatment is able to inhibit existing pit growth and prevent initiation of any new pits; and therein to conclude that the pitting had occurred during exposure to hydrotest and/or raw sea water and had not grown once exposed to inhibited produced fluids – i.e. after field start-up.

Flexible risers, due to their development and ready availability, now used as a mix, make-up configuration or wholly alone, are a critical technology in advancing the use of floating production storage and offloading production vessels (FPSOs). This has arguably been something of game changer technology. However, while their construction is such that internal corrosion is not generally viewed as a credible threat, they do require careful handling during laying and hook-up as the outer polymer sheath can be easily damaged. Access of raw sea water into the annulus housing the tensile wires (providing bending strength) can drastically reduce a riser's fatigue life – from essentially infinity to a matter of months – if immediate remedial action is not undertaken. Severe flooding of the annulus may also seriously compromise the wire layer providing pressure containment. Clamping damaged locations in the outer polymer sheath together with flushing and filling the annulus with an inhibited fluid to displace the raw sea water can return a safe and operationally acceptable fatigue life – 10+ years if addressed quickly enough and depending on the service of the riser – otherwise a replacement riser is the only sure answer. Undertaking meaningful inspection of a flexible riser is a major challenge given its composite structure.

19.5 Summary

For the foreseeable future fossil fuel will remain a primary source of energy with substantial growth in gas to meet the growing demand. In delivering the anticipated growth in energy demand, the search for oil and gas has moved to deepwater, remote areas and HPHT reservoirs. In exploring and producing from such reservoirs, technology plays a significant role. Materials development, corrosion control measures, and recognition of credible threats with appropriate risk assessments are key avenues affecting the majority of these discipline area. The sector considers effective deployment of right technology as being key to business success with a phased implementation of step-outs, game changes, and innovative avenues.

The corrosion and material discipline is considered a vital constituent in hydrocarbon production. Innovations made over the years have contributed significantly to the provision of cost optimisation, safety, and security while adhering to increasingly stringent environmental challenges. Nevertheless, it is shown that implementation of innovative solution needs to be taken on board with great caution and in particular due care in metallurgical processes and QA/QC being implemented effectively and consistently. In addition, for materials selection, a need for rigorous evaluation of transient as well as steady state environmental conditions are vital elements of a methodical strategy to materials optimisation.

References

1 BP, BP technology outlook 2035, 2016.
2 BP, BP energy outlook 2035, 2016

3 IEA, World energy outlook, 2014.

4 ExxonMobil Corporation, The outlook for energy: a view to 2040, 2017.

5 Belani, A. and Orr, S. (2008). A systematic approach to hostile environments. *Journal of Petroleum Technology* 60 (7): 34–39.

6 AW Glass (ed.), High pressure, high temperature developments in the United Kingdom Continental Shelf UK, Report 409, HSE, 2005.

7 Avant, C., Behera, B.K., Danpanich, S. et al. (2012). Testing the limits in extreme well conditions. *Oilfield Review* 24 (3): 4–19, Autumn.

8 de Bruijn, G., Greenaway, R., Harrison, D. et al. (2008). High pressure, high temperature technologies, *Oilfield Review* 46–60, Autumn.

9 Department of Trade and Industry, Trade Partners UK. 2017. www.tradepartners.gov. uk/oilandgas/profile/index/worldmarket.shtml (accessed 8 October 2018).

10 Douglas Westwood Ltd, Various reports (http://www.dw-1.com), 2001–2003 (accessed 8 October 2018).

11 NACE, International measures of prevention, application and economics of corrosion technologies study (IMPACT), Report No. OAPUS310GKOCH (PP110272)-1, 2016.

12 I M Hannah and D Seymour, Shearwater super duplex tubing failure investigation. NACE Annual Corrosion Conference, Paper No. 06491, 2006.

13 N Renton, D Seymour, I Hannah, and W Hughson, A new method of material categorisation for super-duplex stainless steel tubulars. SPE High Pressure/High Temperature Sour Well Design Applied Technology Workshop, SPE 97591, 2005.

14 P J Webb, The role of electrolyte simulation in understanding the failure of shearwater process pipework. SPE International Symposium on Oilfield Corrosion, 28 May, Aberdeen, SPE 87559-MS, 2004.

15 S Huizinga, B McLaughlin, W E Leik, and J G de Jong, Offshore nickel alloy tubing hanger and duplex stainless steel piping failure investigations. NACE Annual Corrosion Conference, Paper No. 03129, 2003.

16 B McLoughlin, S Huizinga, J G De Jong, et al., Offshore 22Cr duplex stainless steel cracking – failure and prevention. NACE Annual Corrosion Conference, Paper No. 05474, 2005.

17 Sentence, P. (1991). Hydrogen embrittlement of cold worked duplex stainless steel oilfield tubulars. In: *Duplex Stainless Steels '91*, vol. 2, 895–903. Les Editions de Physique,.

18 T Cassagne, M Bonis, D Hillis, et al., Understanding field failures of alloy 718 forging materials in HPHT wells. EuroCorr, 2008.

19 S K Mannan, E L Hibner, and B C Puckett, Physical metallurgy of alloys 718, 925, 725, and 725HS for service in aggressive corrosion environments. NACE Annual Corrosion Conference, Paper No. 03126, 2003.

20 S Shademan, J W Martin, and A P Davis, USNS N07725 nickel alloy failure. NACE Annual Conference, Paper No. C2012–0001095, 2012.

21 T Cassagne, M Bonis, C Duret, and JL Crolet, Limitations of 17–4 PH metallurgical, mechanical and corrosion aspects. NACE Annual Corrosion Conference, Paper No. 03102, 2003.

22 G Emygdio and A Zeemann, Failures of 17–4 PH steel parts in non sour environments, NACE Annual Corrosion Conference, Paper No. 4298, 2014.

23 R Mack, Intergranular fracture of UNS 17400, a precipitation hardenable, corrosion resistant alloy used for E&P service. NACE Annual Corrosion Conference, Paper No. 02065, 2002.

24 SM Hesjevik, S Olksen, and G Rorvik, Hydrogen embrittlement from cathodic protection on supermartensitic stainless steels – a case history. NACE Annual Conference, Paper No. 04545, 2004.

25 D Ray, Corrosion resistant alloys for flowlines, manifolds and pipelines; status. BP, 4th ADCO International Corrosion Conference, Abu Dhabi, February 2004.

Bibliography

Royal Dutch Shell, LNG Outlook, 2017.

Abbreviations

AC	alternating current
AEP	annual engineering plan
AIM	asset integrity management
APR	annual performance review
bbl	barrel
BoD	basis of design
boe	barrel of oil equivalent
BOL	bottom of line
CA	corrosion allowance
CAPEX	capital expenditure
CCS	carbon capture and storage
CCT	critical crevice temperature
CCUS	carbon capture use and storage
CDF	critical dilution factor
CE	carbon equivalent
CFD	computational fluid dynamics
CH	cold hardening
CI	corrosion inhibitor
CIP	close interval potential
Cl⁻SCC	chloride stress corrosion cracking
CLAS	carbon and low alloy steel
CP	cathodic protection
CPET	corrosion protection evaluation tool
CPT	critical pitting temperature
CRA	corrosion resistant alloy
CUI	corrosion under insulation
CW	cold working
DAA	D-amino acids
DC	direct current
DCVG	direct current voltage gradient
DMR	direct magnetic response
DMS	data management system
DP	dew point

Corrosion and Materials in Hydrocarbon Production: A Compendium of Operational and Engineering Aspects, First Edition. Bijan Kermani and Don Harrop.
© 2019 John Wiley & Sons Ltd. This Work is a co-publication between John Wiley & Sons Ltd and ASME Press.

DPT	dye penetrant testing
DSS	duplex stainless steel
D2D	data-to-desk
E&P	exploration & production
EC	environmental cracking
ECI	electrochemical impedance
E-Cl⁻SCC	external chloride SCC
ECN	electrochemical noise
EFC	European Federation of Corrosion
EOR	enhanced oil recovery
ER	electrical resistance
ESS	expandable sand screen
FBE	fusion-bonded epoxy
FE	finite element
FEA	finite element analysis
FEMA	failure modes and effects analysis
FFKM	perfluoroelastomers
FFP	fit for purpose
FFS	fit for service
FIMS	facilities integrity management system
FIPP	foamed in place polyurethane
FKM	fluorocarbon rubber
FPSO	floating production storage and offloading
FRP	fibreglass reinforced plastic
FSM	field signature method/monitoring
GEO	gas engine oil
GHSC	galvanically induced hydrogen stress cracking
GIS	geographical information system
GLR	gas liquid ratio
GnAB	gram-negative anaerobic bacteria
GOR	gas oil ratio
GPS	global positioning systems
GRE	glass reinforced epoxy
HAC	hydrogen assisted cracking
HAZ	heat-affected zone
HAZID	hazard identification
HAZOP	hazard and operability
HDPE	high density polyethylene
HE	hydrogen embrittlement
HIC	hydrogen induced cracking
HIPO	high potential incidents
HNBR	hydrogenated nitrile
HP	high pressure
HPHT	high-pressure high temperature
HS&E	health, safety and environment
ICP	inductively coupled plasma
ILI	in-line inspection
KPI	key performance indicator

LOPA	layer of protection analysis
LPR	linear polarisation resistance
LTO	licence to operate
MAOP	maximum allowable operating pressure
MAWT	minimum allowable wall thickness
MDEA	methyl-di-ethanol-amine
MDPE	medium density polyethylene
MEG	monoethylene glycol
MFL	magnetic flux leakage
MIC	microbial induced corrosion
MPT	magnetic particle testing
MSS	martensitic stainless steel
MT	magnetic particle testing
NBR	nitrile rubber
NDE	non-destructive evaluation
NETL	National Energy Technology Laboratory
NORM	naturally occurring radioactive material
NPS	nominal pipe size
OCP	open circuit potential
OPEX	operating expenditure
OS	oxygen scavenger
P	pressure
P&ID	piping and instrumentation diagram
PA	polyamide
PCM	pipeline corrosion management
PCT	plastic-coated tubular
PE	polyethylene
PEC	pulsed eddy current
PEEK	polyether-ether-ketone
PE-RT	polyethylene raised temperature
PFD	process flow diagram
PH	precipitation hardening
PHA	process hazard analysis
PI	performance indicators
PIMS	pipeline integrity management system
PMS	pipeline management system
POF	probability of failure
PP	polypropylene
ppb	parts per billion
ppm	parts per million
PPS	polyphenylene sulphide
PREN	pitting resistance equivalence number
PT	penetrant testing
PW	produced water
PWC	preferential weld corrosion
PWHT	post-weld heat treatment
Q&T	quenching and tempering
QA	quality assurance

QC	quality control
QPR	quarterly performance review
QRA	quantitative risk assessment
RAM	reliability, availability and maintainability
RBA	risk-based assessment
RBI	risk-based inspection
RDS	rate-determining step
ROV	remote operating vehicle
RPCM	ring pair corrosion monitoring
RT	radiography testing
RTP	reinforced thermoplastic pipes
SA	solution annealed
SCC	stress corrosion cracking
SCE	safety critical equipment
SCR	steel catenary riser
SDSS	super duplex stainless steel
SIC	shallow internal corrosion
SLC	sustained load cracking
SME	subject matter expert
SMSS	super MSS
SOHIC	stress-oriented hydrogen induced cracking
SRB	sulphate reducing bacteria
SS	stainless steel
SSC	sulphide stress cracking
SW	sea water
SWC	stepwise cracking
SZC	soft zone cracking
T	temperature
TDS	total dissolved solid
TFEP	tetrafluoroethylene-propylene rubber
THPS	tetrakis (hydroxymethyl) phosphonium sulphate
TLC	top-of-line corrosion
TMCP	thermo-mechanically controlled processes
TOL	top of line
TOLC	top-of-line corrosion
TRB	thiosulphate-reducing bacteria
TSA	thermally sprayed aluminium
UT	ultrasonic testing
UTS	ultimate tensile strength
VCI	vapour phase corrosion inhibitors
VOC	volatile organic compounds
VT	visual testing
WAG	water alternating gas
WI	water Injection
WIMS	wells integrity management systems
WLC	whole life cost

Index

*Corrosion and Materials in Hydrocarbon Production: A Compendium of Operational
and Engineering Aspects*, First Edition. Bijan Kermani and Don Harrop.
© 2019 John Wiley & Sons Ltd. This Work is a co-publication between John Wiley & Sons Ltd and ASME Press.